MONOLAYER
AND SUBMONOLAYER
HELIUM FILMS

MONOLAYER AND SUBMONOLAYER HELIUM FILMS

Edited by
John G. Daunt and E. Lerner

Cryogenics Center
Stevens Institute of Technology
Hoboken, New Jersey

PLENUM PRESS · NEW YORK-LONDON

Library of Congress Cataloging in Publication Data

Symposium on Monolayer and Submonolayer Helium Films,
  Stevens Institute of Technology, 1973.
  Monolayer and submonolayer helium films.

  Includes bibliographical references.
  1. Thin films – Congresses. 2. Helium at low temperatures – Congresses. I. Daunt,
John Gilbert, 1913–   ed. II. Lerner, Eugenio, ed. III. Title.
QC176.82.S9 1973   530.4'1   73-12930
ISBN-13: 978-1-4613-4582-4   e-ISBN-13: 978-1-4613-4580-0
DOI: 10.1007/ 978-1-4613-4580-0

Proceedings of the Symposium on Monolayer and Submonolayer
Helium Films held at the Stevens Institute of Technology,
Hoboken, New Jersey, June 7-8, 1973.

© 1973 Plenum Press, New York
Softcover reprint of the hardcover 1st edition 1973
A Division of Plenum Publishing Corporation
227 West 17th Street, New York, N. Y. 10011

United Kingdom edition published by Plenum Press, London
A Division of Plenum Publishing Company, Ltd.
Davis House (4th Floor), 8 Scrubs Lane, Harlesden, London, NW10 6SE, England

EDITORIAL FOREWORD

This volume is devoted to the topic of "Monolayer and Sub-monolayer Helium Films," which was the subject of a symposium held at Stevens Institute of Technology, June 7th and 8th, 1973. All the papers in this volume were presented at this symposium. The symposium was sponsored by Stevens Institute of Technology with support from the National Science Foundation and the Office of Naval Research, and had as Organizing Committee: Professor D. F. Brewer (University of Sussex), Professor J. G. Dash (University of Washington), Professor J. G. Daunt (Stevens Institute of Technology), Doctor V. J. Emery (Brookhaven National Laboratory), Professor E. Lerner (Stevens Institute of Technology), Doctor F. J. Milford (Battelle Memorial Institute), Doctor A. D. Novaco (Brookhaven National Laboratory), Professor F. Pollock (Stevens Institute of Technology), and Professor W. A. Steele (Pennsylvania State University).

The symposium was largely devoted to papers on the topic of thin (unsaturated) helium films at low temperatures, many of which were of a review character. In addition some papers of a more general character were included in order to review some powerful techniques used for investigating surfaces at higher temperatures. The Editors wish to thank all the authors of these papers for their enthusiastic cooperation in the preparation of this volume.

John G. Daunt
E. Lerner
Editors

CONTENTS

PHASES OF HE[3] MONOLAYER FILMS ADSORBED ON GRAFOIL[†]

Susanne V. Hering and Oscar E. Vilches

Department of Physics, University of Washington

Seattle, Washington 98195

I INTRODUCTION

Monolayer films of helium atoms adsorbed on Grafoil (1) show a variety of different phases that have made their study attractive to both theorists and experimenters. Following the experimental discovery of a quantum gas phase by Bretz and Dash (2), the same authors found at higher densities a substrate ordered phase (3), and a He[4] two dimensional (2D) solid phase (4). Only a few He[3] coverages were investigated in the above experiments. A more detailed study of He[3] was carried out by McLean (5) over a larger temperature range (.036K< T < 4.2K) but only over a limited density range (.25 < x < .62; x ≡ partial monolayer coverage). A full report on all the above experiments will be published shortly. (6)

This paper summarizes the present knowledge of the behavior of He[3] monolayer films adsorbed on Grafoil. It includes the measurements reported in References 5 and 6, plus new heat capacity measurements at coverages lower than x = .25 and larger than x = .62.

II EXPERIMENTAL

The experimental set up has been described extensively in Ref. 6. The cell used in all the new He[3] experiments reported here is the one called "Cell B" in Ref. 6. For that cell, one monolayer of He[3] corresponds to 108.6 cm[3] STP of gas. The area of the adsorbent is measured using the lattice gas ordering transition of He[4], which gives a large heat capacity peak when the fraction of graphite hexagons occupied by a helium atom is $x_g = 1/3$. This method yields an areal density at one monolayer of n = .110 atoms per square angstrom.

III EXPERIMENTAL RESULTS

a. Very Low Coverages

The lowest investigated coverage is x ≃ .02. Between this coverage and about a tenth of a layer, results are difficult to interpret. Although heat capacity signals at very low temperature resemble 2D solids, at temperatures of the order of three or four degrees they increase monotonically exceeding the classical value of 2k per atom. The x = .02 gives C/Nk ≃ 3 at T = 4°K, while the x ≃ .05 gives C/Nk ≃ 2.4 at the same temperature. The last coverage is represented by a dashed line in Fig. 1 rather than by the

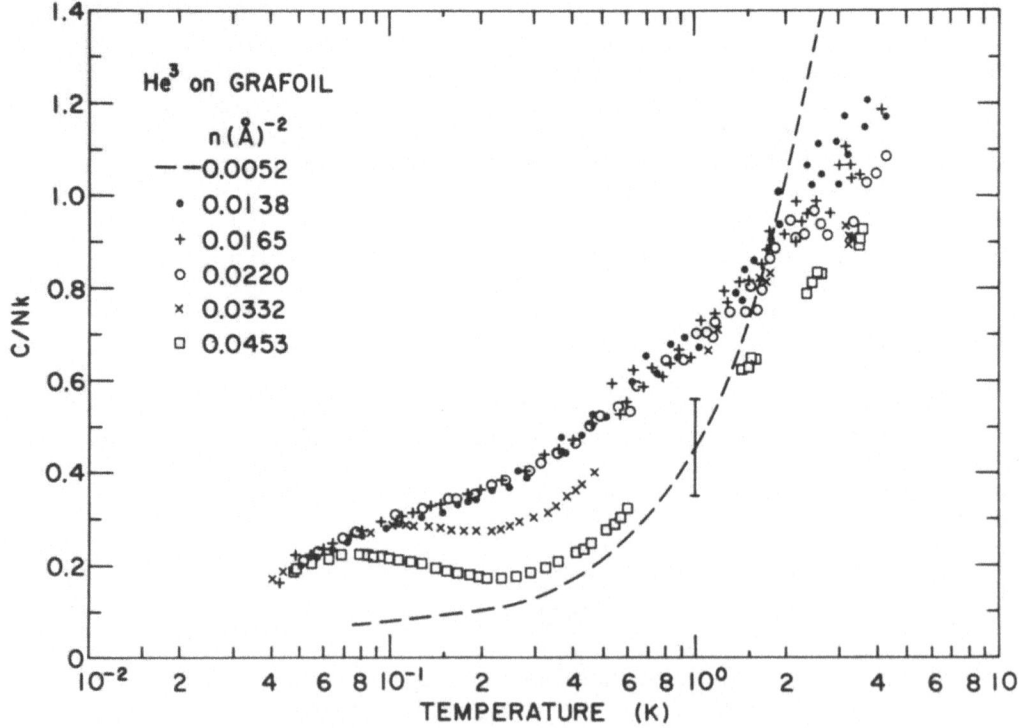

Fig. 1 The specific heat of low coverage He³ films adsorbed on Grafoil.

actual experimental points, since the scatter of the heat capacity measurement is large for T greater than about .8K. This is due to the helium signal being only 30 to 50% of the small total heat capacity. Furthermore, results are somewhat irreproducible. Annealing of the samples at a temperature well above $4°K$ for several hours produces adsorbed films of different type, judging by variations of the measured specific heat. This effect is not observed if the temperature is never raised above four or five degrees or at higher coverages, where after the first annealing, successive annealings do not change the specific heat.

Bretz (7) observed similar behavior for He^4 films on Grafoil. His measurements gave C/Nk ≈ 1.8 and 1.2 at T = $4°K$ for x ≈ .05 and .1 respectively. The low temperature portion of his results could be fitted by a 2D Debye solid of θ_D ≈ $5.5°K$.

Elgin (8) has measured the heat capacity of several low coverage He^4 films and interpreted his and Bretz's data as an inhomogeneity compressed phase. The inhomogeneities occur in the regions where two adsorbing graphite planes join at small angles. There the binding energy can be substantially increased in a more uniform way. The large heat capacity signals at higher temperatures are due to helium atoms "evaporating" from the compressed region to a higher entropy 2D gas phase. The fraction of the total area considered non uniform that best fits the data is about 2%.

b. Low Coverages

For coverages greater than x ≈ .15 up to x ≈ .52 He^3 films behave close to 2D ideal gases. For T > $2°K$ the specific heat reaches values near 1k per atom, close to the classical 2D gas value. The specific heat decreases towards zero as T drops below $1°K$, almost like a 2D ideal quantum gas. There is no trace of the one degree peaks observed at equivalent coverages for He^4 (2). For temperatures lower than $.5°K$, the specific heat deviates markedly from the one of the 2D gas, and goes through a broad peak of maximum value at T from about $.12°K$ for the lowest coverage to $.07°K$ for the higher ones.

Three more coverages besides those reported in Refs. 5 and 6 have been measured. They are shown in Fig. 1 at n = .0138, .0165 and .0220 $Å^{-2}$ (or x ≈ .125, .15 and .20). The high temperature scatter of the data is a consequence of the experiment being actually designed to investigate the onset of the low temperature anomaly. Still it is easy to see that as coverage increases the asymptotic value of C/Nk decreases. Values above one for the lowest coverages probably include a contribution from the very low coverage solid. For the intermediate and higher coverages reported

in Refs. 5 and 6, $(C/Nk)_{4^\circ K} < 1$; this has been recently ascribed
(9) to two body interactions between the He^3 atoms that result in
a contribution to the specific heat from the second derivative of
the second virial coefficient. Furthermore, this calculation also
predicts that at low temperature the specific heat will begin to
increase, the temperature of the rise being remarkably close to
that at which the experimental heat capacity increases. Still
though the actual nature of what occurs is not well understood.
Entropy calculations for similar coverages of He^3 and He^4 done by
extrapolating to T = 0 the C vs. T curves and integrating C/T as
a function of T show that a considerable amount of the He^3 spin
entropy is developed through this anomaly (6). The actual low
temperature joining of the specific heat curves even to quite high
coverages (see Fig. 1) could signal the beginning of a 2D liquid
state.

c. Ordered Lattice Gas

For coverages at least between $.061 \overset{\sim}{\lesssim} n \overset{\sim}{\lesssim} .070 Å^{-2}$ large peaks
develop at T around $3^\circ K$. Above the peak the signal falls to that
of a 2D gas, while below it drops to zero as $\exp(\alpha T)$. These peaks
are fairly symmetric in T being logarithmic on both sides of the
transition with almost identical coefficients for the transition
occurring at the highest temperature (6). Their maximum value
changes rapidly with coverage. The temperature at which they
occur does not change so rapidly, but the maximum height peak
occurs at the maximum temperature. Results of a survey taken in
steps of about 1% coverage are shown in Fig. 2.

This peak has been interpreted (3) as being due to a transition
from a gas phase to an ordered lattice gas phase in which the He
atoms form an ordered array occupying 1/3 of the adsorbing sites
formed by the carbon hexagons of the graphite. The transition
occurs for both He^3 and He^4, although the He^3 transition seems to
occur at about 1.5% higher coverage. The origin of this discrepancy, seemingly outside the experimental error, is not well understood at present.

The existence of an $x_g = 1/3$ ordered phase has been theoretically
investigated and verified by Campbell and Schick (10) for a classical gas. The fact that the transition occurs for He^3 at $.1^\circ K$
higher temperature than for He^4 can be explained by quantum mechanical differences between the two isotopes (9, 11). Further
evidence for the existance of an ordered array has been obtained
by Rollefson (12) using NMR to study He^3 adsorbed on graphitized
carbon black, a material similar to Grafoil.

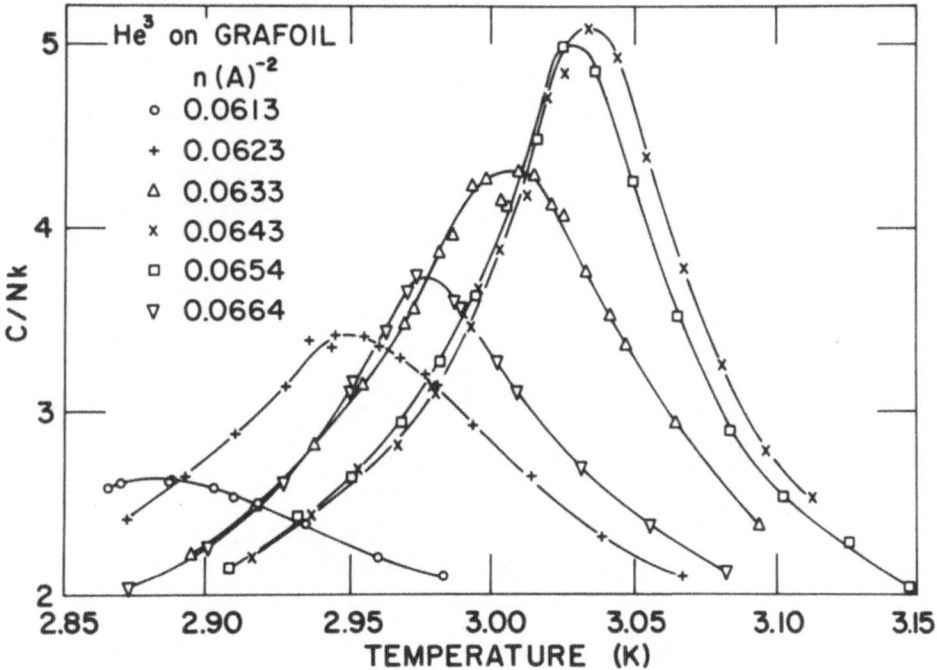

Fig. 2   The specific heat in the vicinity of the lattice gas
         ordering transition.

d. High Coverages

Within a coverage range of only .007 $Å^{-2}$ the nature of the
adsorbed films changes completely. The ordering peak still exists
for n = .070 but it has dropped in height at T_p to C/Nk ≈ 1.17 and
in temperature to 2.88°K. For N = .077$Å^{-2}$ it has disappeared and
a sharp small peak is seen at T ≈ 1.03°K. Increases in coverage
of about 5 cm³ STP of gas produce a steady increase in both the
size of the peak and the temperature at which they occur. The
four coverages that have been investigated to date are shown in
Fig. 3.

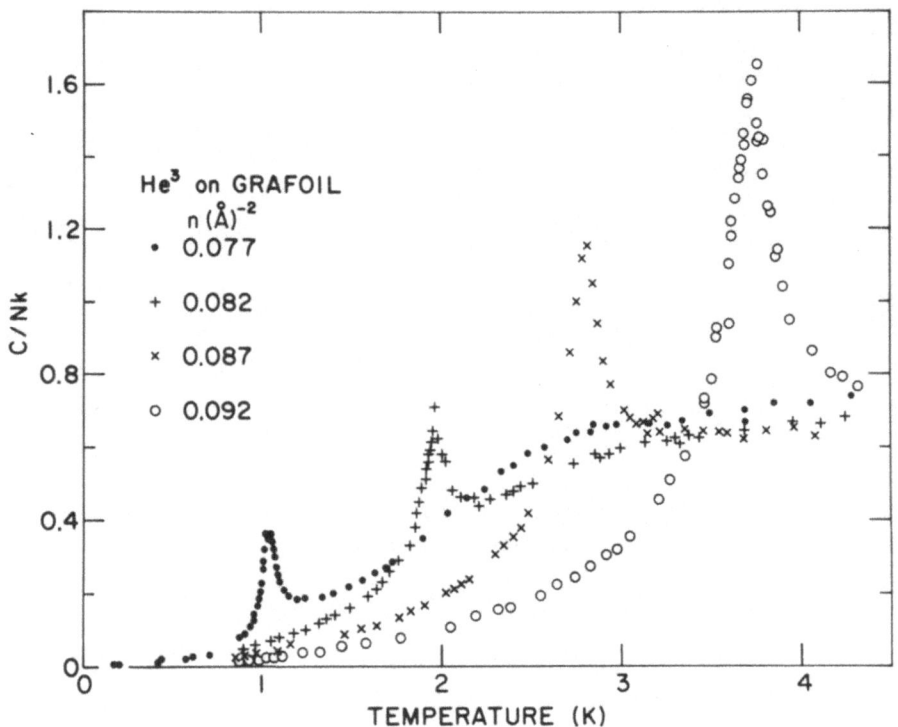

Fig. 3 Specific heat peaks along the melting line

At temperatures well below the peak the specific heat is nearly quadratic in T, as shown in Fig. 4.

   Although four coverages are not enough to draw elaborate conclusions, by analogy to results with He^4 films of Bretz et. al. (4) and based on the T^2 low temperature specific heat, we conclude that the peaks are indicative of the formation of a 2D solid. So far it seems that the melting temperatures are not too different for both isotopes, but the close correspondence found between the

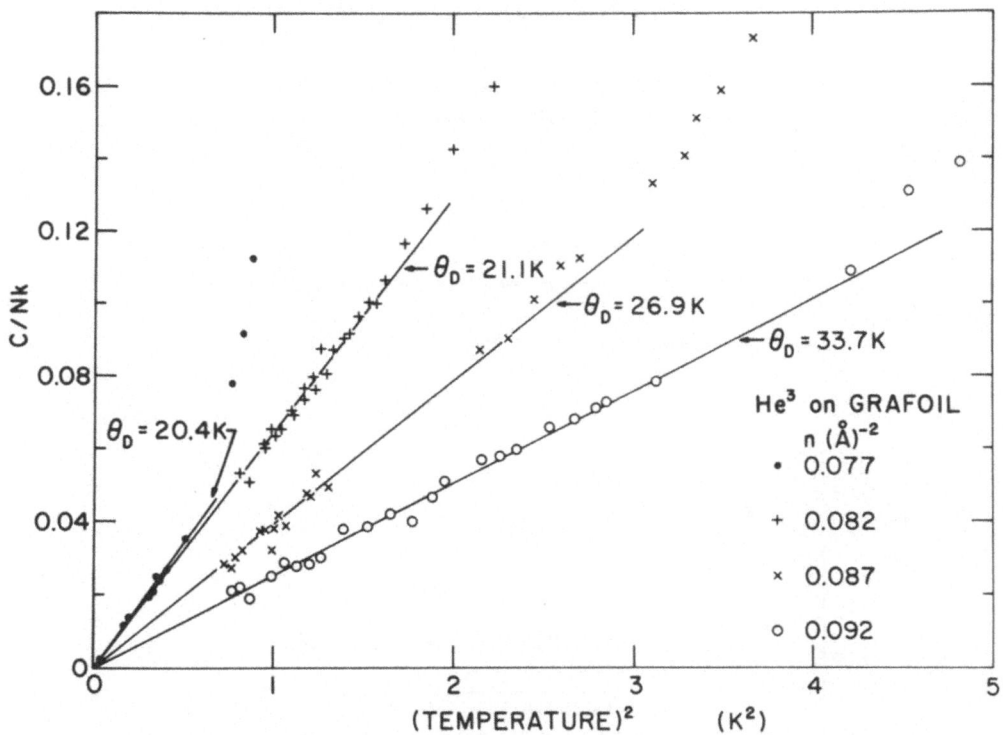

Fig. 4   The low temperature specific heat versus T^2 for the
2D solid. Characteristic temperatures calculated using
2D Debye model.

Debye temperatures of hcp He[4] and the ones of the films is not
found in the case of He[3]. Scaling densities from 2 to 3 dimensions
for the coverage range investigated in Figs. 3 and 4 makes the
film data correspond to the bcc phase (13) of He[3]. Differences
in θ are then of the order of 7°K, being higher for the film.

Bretz and Dash (14) have studied the possibility of continuous
melting of a 2D solid of finite size on a mode by mode basis.
Melting is equivalent to the decrease with temperature of extended
short range order, small wavevector modes being the first ones

to melt. Their simple model actually gives a melting temperature independent of the mass of the particles forming the solid.

IV THE PHASE DIAGRAM

All the regions investigated so far have been indicated in the areal density vs. temperature diagram of Fig. 5.

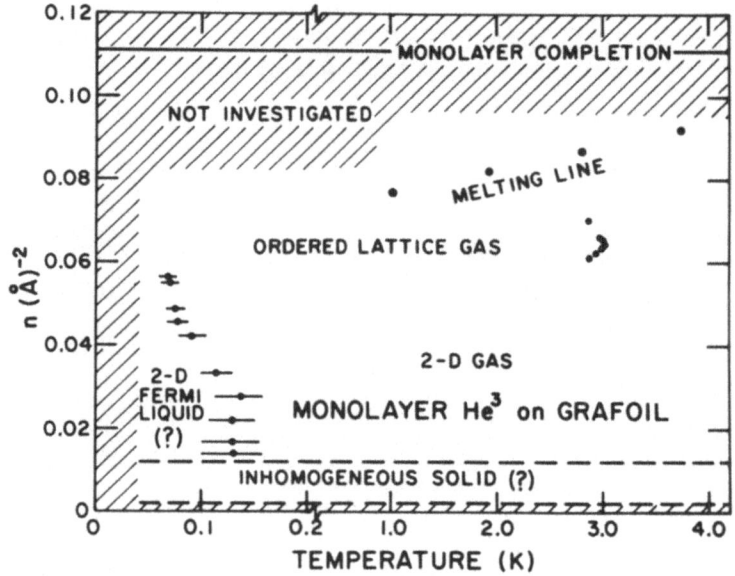

Fig. 5   Phase diagram of 2D He^3 adsorbed on Grafoil.

No experiments have been done below $n \simeq .022$ or above $n \simeq .092 Å^{-2}$, except for one coverage at almost monolayer completion taken by Bretz et al. (4) to verify that the peaks they were observing were not just due to He^4 (and perhaps superfluidity). Also no experiments have been done for $T \lesssim .04°K$ or $T \gtrsim 4.2°K$. For that matter, no experiments have been done at coverages between those indicated by the points of Fig. 5. Of particular interest to us is the region between the lattice gas ordering transition and the solid. The highest coverage that we have investigated so far for which there is an indication of the $x_g = 1/3$ transition shows no other anomaly in C/Nk down to .1°K. There still is a .07 mono-layer coverage range to investigate to find where the new solid phase appears.

Measurements along the melting line should continue to higher coverages to complete the comparison with both He^4 film measurements and solid He^3. If promotion to the second layer is responsible for the very large specific heat peaks near monolayer completion (8) one would expect differences in coverages and temperatures at which the increase begins to show.

The nature of the low temperature transition probably can not be completely decided on the basis of heat capacity experiments alone, but measurements to much lower temperature should decide whether all coverages give the same specific heat. The fact that entropies at $4°K$ calculated from the measured specific heats are higher than those of He^4 for coverages which show the low temperature anomaly and smaller for those of the ordered phases with no low T anomaly could be studied with vapor pressure measurements analogous to those done by Elgin (8) in He^4. Ideal gas theory predicts that the He^3 entropy should always be higher than the He^4 entropy at temperatures sufficiently high that the theory applies. If this was verified by vapor pressure measurements, then one could expect a very low temperature spin anomaly in the solid phase.

REFERENCES

†Work supported by the National Science Foundation.
1. Grafoil is a product marketed by Union Carbide, made with natural Madagascar graphite.
2. M. Bretz and J. G. Dash, Phys. Rev. Letters 26, 963 (1971).
3. M. Bretz and J. G. Dash, Phys. Rev. Letters 27, 647 (1971).
4. M. Bretz, G. B. Huff, and J. G. Dash, Phys. Rev. Letters 28, 729 (1972).
5. E. O. McLean, Ph.D. Thesis, U. of Washington (1972) (unpublished). A partial description of the very low temperature results was published, D. C. Hickernell, E. O. McLean and O. E. Vilches, Phys. Rev. Letters 28, 789 (1972).
6. M. Bretz, J. G. Dash, D. C. Hickernell, E. O. Mclean and O. E. Vilches, to be published in The Physical Review.
7. M. Bretz, Ph.D. Thesis, U. of Washington (1971) (unpublished).
8. R. L. Elgin, Ph.D. Thesis, CalTech (1973) (unpublished).
9. R. Siddon, Ph.D. Thesis, U. of Washington (1973) (unpublished). See also R. Siddon and M. Schick, this conference.
10. C. E. Campbell and M. Schick, Phys. Rev. A5, 1919 (1972).
11. R. Siddon and M. Schick, 13th International Conference on Low Temp. Physics, Boulder, (1973) (to be published).
12. R. J. Rollefson, Phys. Rev. Letters 29, 410 (1972) and this conference.
13. H. H. Sample and C. A. Swenson, Phys. Rev. 158, 188 (1967).
14. M. Bretz and J. G. Dash, J. Low Temp. Physics 9, 291 (1972).

SPECIFIC HEAT OF 2ND LAYER HE^4 FILMS[*]

Michael Bretz[†]

Department of Physics, University of Washington

Seattle, Washington 98195

Today I will present some recent heat capacity results on the 2nd monolayer of He^4 adsorbed on graphite. One might ask why study the 2nd layer, of what importance is it? Well, I believe that a comparison of phenomena occurring on the 2nd and 1st layers will confirm some of the interpretations of monolayer behavior and will lend evidence in those regimes where the interpretations are still in doubt. But further, the 2nd layer appears in some ways a superior system for investigation of 2-Dimensional film phenomena thus making it an interesting system in its own right.

Let's consider the substrate for 2nd layer adsorption - that is, the He^4 complete monolayer. We have just seen that near completion the monolayer becomes a dense solid possessing high θ Debye's. This 2D solid is known to be compressible (1) and an observed shift of the melting peak from 7.38K for the complete but bare monolayer to 8.62K in the presence of a partial 2nd layer reflects an increase in density of the monolayer. But more to the point, it illustrates the continued existence and relative stability of the monolayer solid when further layers are added. Second layer atoms should see weaker wells of different configuration than do atoms in the monolayer. We expect substrate inhomogeneities to be smoothed over by the monolayer solid resulting in a more uniform, substrate for 2nd layer adsorption than is bare graphite.

From 4.2 and 2K isotherms we determine that the monolayer and 2nd layer complete at about 96.5 and 165 cc STP respectively, and that the 2nd layer isosteric heat of adsorption is in the range 50-30K. In all the results presented here the solid monolayer

11

and empty calorimeter heat capacity contributions have been subtracted out (2).   Excess heat capacity due to desorption has been
determined according to a thermodynamic relation between pressure
and desorption heat capacity (3) and has also been subtracted from
the data.   In determining the areal density for each coverage it
is assumed that the 2nd layer possesses the same adsorption area
as the monolayer.

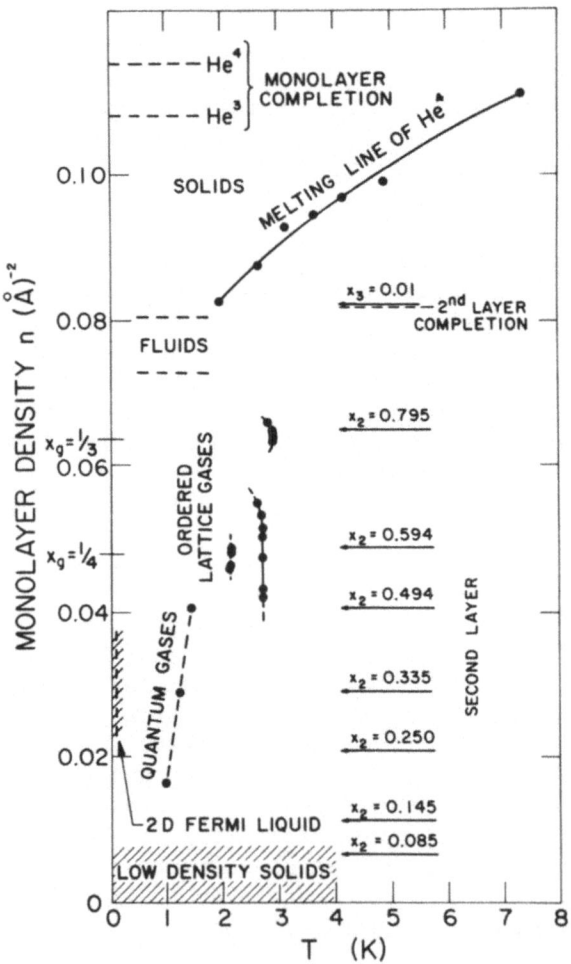

Fig. 1   Phase diagram of monolayer He[4] with arrows indicating the
         equivalent densities of runs on the 2nd layer.   These are
         presented in Fig. 2. x_2 is partial 2nd layer coverage,
         where $x_2 = 1$ means 165 STPcc.

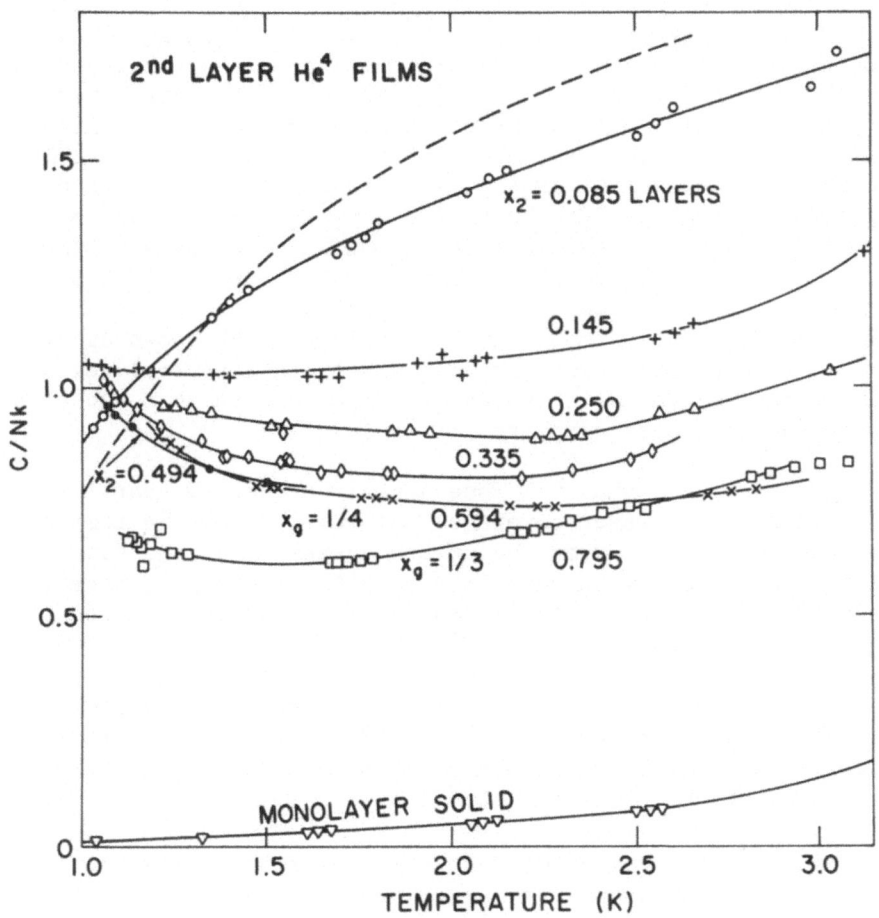

Fig. 2 He^4 specific heat of coverages on the 2nd layer. Dotted line gives 1st layer results for n = .00624Å^{-2}.

Figure 1 is the familiar monolayer phase diagram. 8 runs were taken on the 2nd layer which correspond to the areal densities of the interesting monolayer phase regions. The densities of these runs are indicated by the arrows. Figure 2 contains most of the second layer runs which are identified as fractional 2nd layer coverage x_2. Notice that the monolayer specific heat contribution is quite small so any changes in θ Debye of the monolayer due to 2nd layer adsorption will make little difference in the data analysis. Notice also that even after correcting for desorption high temperature heat capacity tails are still present.

These are probably due to layer promotion, as desorption implies a
finite occupancy of the upper film layers. For the lowest coverage x_2 = .085 the specific heat is far above $c/_{Nk}$ = 1, agreeing
with the signal of the very low density monolayer (2) (shown by
the dotted line in Fig. 2). This suggests that substrate hetero-
geneity, which should be less for the 2nd layer, is of only minor
importance in determining the monolayer film behavior at these low
densities. The other curves of Figure 2 clearly shows the trend of
generally decreasing specific heat with coverage, analogous to
that found for high temperature partial monolayer signals. Other
than this the curves are more striking for what they don't show
than for what they do. The x_g = 1/4 and 1/3 ordering peaks are
gone completely,but more important, no ordering peak shows up at
any coverage. This indicates very weak substrate potential wells.
Thus, the monolayer He solid appears to be a more uniform substrate
for adsorption than bare graphite.

A big surprise is that the Bose anomaly occurring near 1.2 -
1.4K for moderate monolayer coverages has shifted down in temperature for the 2nd layer. From the upward curvature of the data
near 1K it can be said with certainty that the anomaly is there.
Unfortunately, data was taken only above 1K and so observation of
the peak shape was not possible. The ordered phases on the monolayer interfere with the anomaly at higher densities, but on the
2nd layer there are no ordered phases. The rise, then fall of the
specific heat with coverage at 1K reflects a corresponding rise
and fall in the temperature of the Bose anomaly, giving a maximum
temperature of maybe .8K at $x_2 \backsim$ 0.6.

Moving on to the region of 2nd layer completion, Fig. 3, a
distinct bump with an anomously large high temperature tail is
observed at x_3 = .01. Both have disappeared again by x_3 = .09.
The tail is consistent with layer promotion, but the bump is
probably not a layer promotion effect. An initial guess might be
that the peak is condensation into the 2D liquid phase, but we
expect a latent heat discontinuity at liquifaction. A plausible
alternative explanation suggested by Greg Dash is that the bump
in 2nd layer solidification occurring just at layer completion.
As the 3rd layer begins forming, interlayer interactions weaken the
solid sufficiently to dissolve it. In any case, by x_3 = .19 a
strong melting peak does emerge. Fig. 4 follows this peak to 3rd
layer completion.

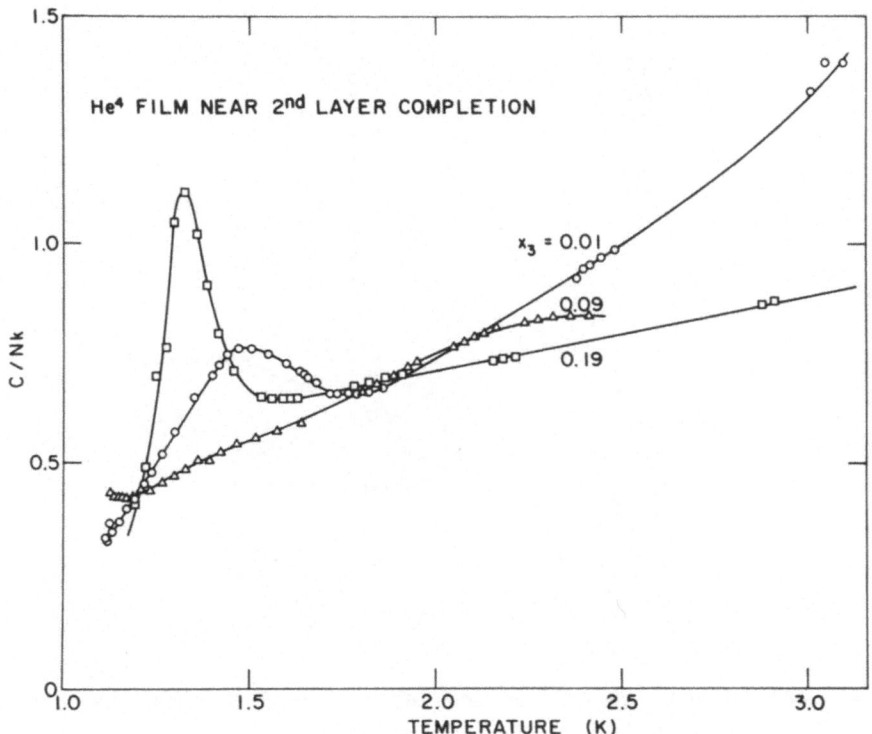

Fig. 3 He⁴ films near 2nd layer completion. x_3 means fractional
3rd layer coverage, taking third (and higher) layer
capacities as 65 STPcc.

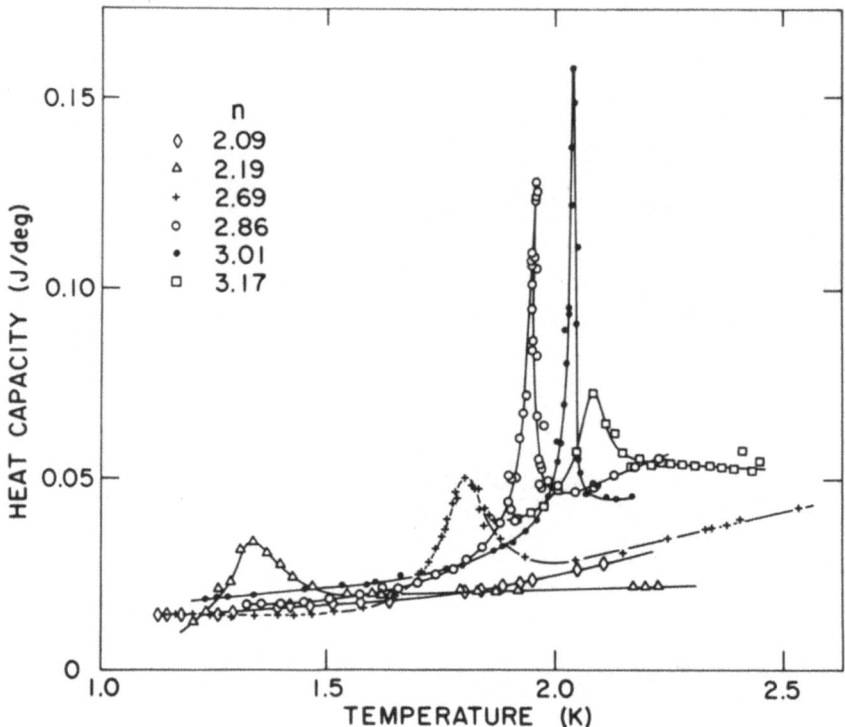

Fig. 4   Heat capacities on the 3rd layer. Here $n = 2 + x_3$ is the film
         thickness in completed atomic layers.

There are a number of reasons why it appears that we are seeing 2nd
layer melting. First, the structure closely resembles the shape
of the monolayer melting peaks. Secondly, they appear at layer
completion where solidification is expected. And thirdly, the
temperature of the peaks fall close to the monolayer melting line.
I feel that the 3rd layer actually causes solidification by
compressing the 2nd layer, just as the monolayer compresses in the
presence of the 2nd layer. In addition to lateral compression the
3rd layer might cause a vertical compression making the 2nd layer
more nearly 2 dimensional and thereby increasing the steric hindrance between atoms. If the solid processes a low temperature T^2
heat capacity dependence it is masked by the gas like contribution

of the third layer atoms. Why the height of the melting peak decreases past 3rd layer completion is not understood at this time.

To summarize, I have constructed a phase diagram of the 2nd layer (Fig. 5) showing the very low density and quantum gas regions, solidification and the melting line. The additional bump at layer completion is indicated by a "?". It is similar to the phase diagram of the monolayer (Fig. 1) except that the ordered phases are missing thus allowing the quantum gases to dominate over most of the 2nd layer. The dotted line gives a rough estimate for the temperature of the Bose anomaly peaks as determined from the 1K heat capacity tails. The melting line is drawn with finite width since the density of the compressed second layer is not known when further layers are adsorbed.

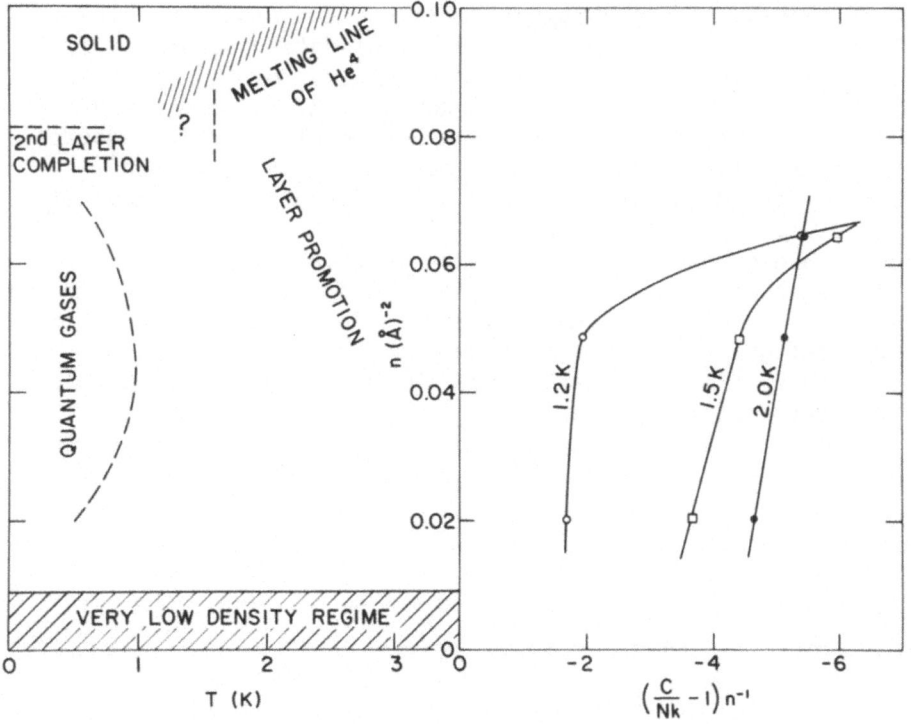

Fig. 5 (left) Phase diagram of the second layer, where n is areal density. (right) Universal specific heat plot of 2nd layer data for different temperatures.

As we'll hear shortly from Bob Siddon and Mike Schick (4), a particularly useful analysis of the quantum gas results is the plot of $(C/Nk - 1)/n$ vs. n or T. To the right of the phase diagram is such a "universal" specific heat plot (5). We see that $(C/Nk - 1)/n$ is nearly a constant, independent of density for low 2nd layer coverages. The slight n dependence indicates the importance of 3-body interactions (3rd virial coefficient). Agreement with the universal plot of the monolayer data is obtained only with a factor of 3 temperature shift, consistent with the temperature shift of the Bose anomaly itself. We see that at higher densities above $x_2 \simeq .6$ this type of analysis breaks down completely.

As a final comment, let me point out that a heat capacity study of 2nd layer He^3 and He^4 below 1K (where layer promotion and desorption are quite small) would be a useful and rewarding project.

REFERENCES

[*]Work supported by the National Science Foundation
[†]Present address: University of Michigan, Ann Arbor, Michigan.
1. G. A. Stewart, S. Siegel, and D. L. Goodstein, Proc. of 13th Int. Conf. on Low Temp. Physics, Boulder, Colo. (Aug. 1972).
2. M. Bretz, J. G. Dash, D. C. Hickernell, E. O. McLean and O. E. Vilches, Phases of He^3 and He^4 Monolayer Films Adsorbed on Basal Plane Graphite, Phys. Rev. A (to be published).
3. J. G. Dash, R. E. Peierls, and G. A. Stewart, Phys. Rev. 2A, 932 (1970).
4. R. L. Siddon and M. Schick, this conference.
5. Coverages of $x_2 = .335$ and $.494$ are not included in the universal plot. This data was acquired quite recently on another cryostat and as of this lecture there remains a small uncertainty in absolute magnitude which would be magnified in this particular analysis.

SOME OBSERVED PROPERTIES OF ^{4}He AND NEON SUBMONOLAYERS[*]

J.G. Daunt, S.G. Hegde and E. Lerner

Cryogenics Center, Stevens Institute of Technology

Hoboken, New Jersey 07030

ABSTRACT

A review is given of recent data obtained for the isosteric heat of adsorption and monolayer coverages from adsorption isotherm measurements of ^{4}He and neon on bare sintered copper and on Grafoil and on these two substrates coated with a monolayer of argon. The data indicate that the argon-coated substrates are more uniform and more homogeneous than the corresponding bare materials. They also show that the isosteric heats of adsorption on the argon-coated materials are rather independent of the base material. A brief review is given of evidences for phase transitions provided by adsorption isotherms and it is shown that only on homogeneous materials such as graphite, Grafoil and alkali-halide crystals do such evidences appear. Data are also provided on the electrical and thermal conductivities at low temperatures of Grafoil.

INTRODUCTION

In the past year detailed measurements have been reported by members[1-10] of the Cryogenics Center on the properties of ^{4}He and neon submonolayers adsorbed on various substrates at low temperatures. It is the

[*]Work supported by a Grant from the National Science Foundation and by contracts with the Office of Naval Research and the Department of Defense(Themis Program)

purpose of this paper to assemble and codify some of the
results of this work, and compare them, where appropriate, with germane data obtained elsewhere. In order to
limit this presentation, only three major sets of data
are presented, namely: (a) Thermodynamic data and
monolayer coverage data, as determined by adsorption
isotherms, (b) data on phase transitions, as determined by adsorption isotherms and (c) transport properties of a major substrate. The data under (a) and
(b) concern ^{4}He and neon submonolayers adsorbed on sintered copper and on Grafoil and on these two substrates
coated with an argon monolayer. The data under (c)
concerns the low temperature transport properties of
Grafoil. Work on specific heats[8,9] and nuclear magnetic resonance[11] of helium submonolayers is not
covered herein.

HEAT OF ADSORPTION AND MONOLAYER COVERAGE

    Detailed measurements of adsorption isotherms[1-4,6,7]
have permitted us to assess the isosteric heat of adsorption and the monolayer coverages of several systems.
Fig. 1 gives the isosteric heat of adsorption, (Q_{st}/R),
at an average temperature, T_{AV}, of about 10K as a func-

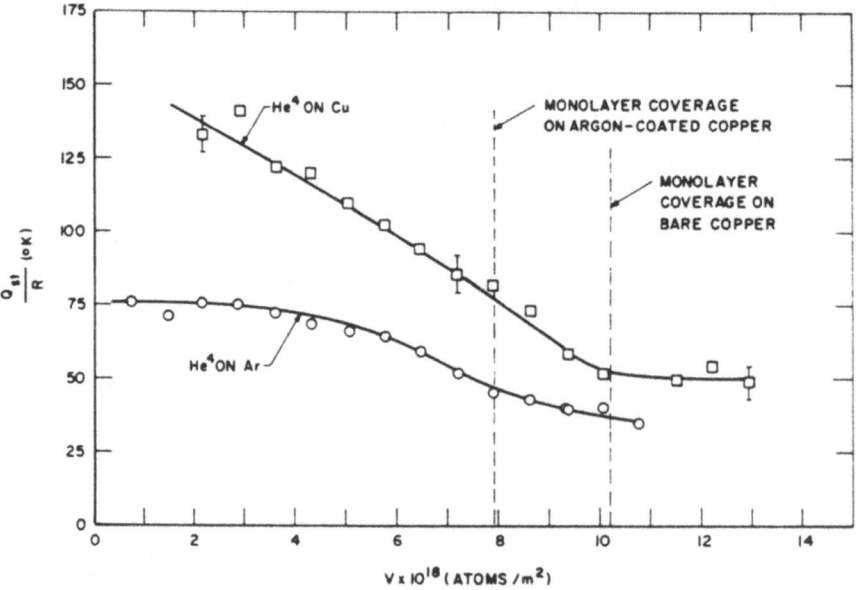

Fig. 1. Isosteric Heat of Adsorption of ^{4}He on bare
copper and on argon-coated copper as a function of
coverage.

tion of coverage for ⁴He on bare sintered copper and on
sintered copper covered with a monolayer of argon[3].
The preparation of the sintered copper followed the
technique reported by Dash et al[12]   In the figure indications are given also of the monolayer coverage on
each adsorbent. Fig. 2 gives similar data for ⁴He on
Grafoil and on Grafoil covered with a monolayer of
argon. Fig. 3 gives similar data for neon adsorbed on
Grafoil covered with a monolayer of argon.

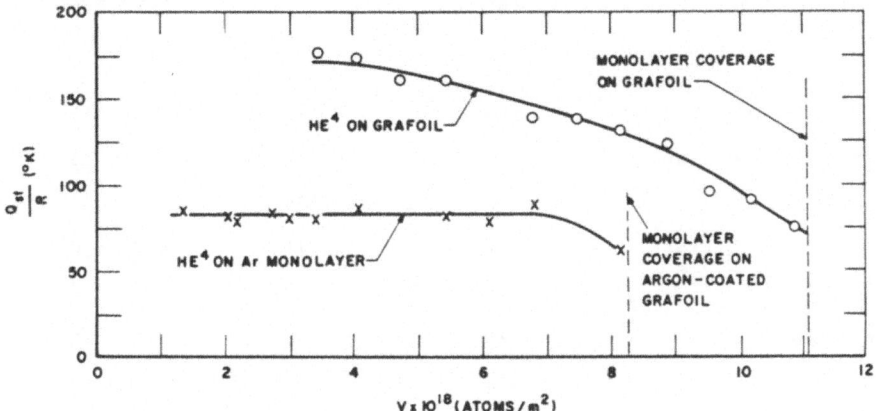

Fig. 2. Isosteric Heats of Adsorption of ⁴He on Grafoil
and on argon-coated Grafoil as a function of coverage.

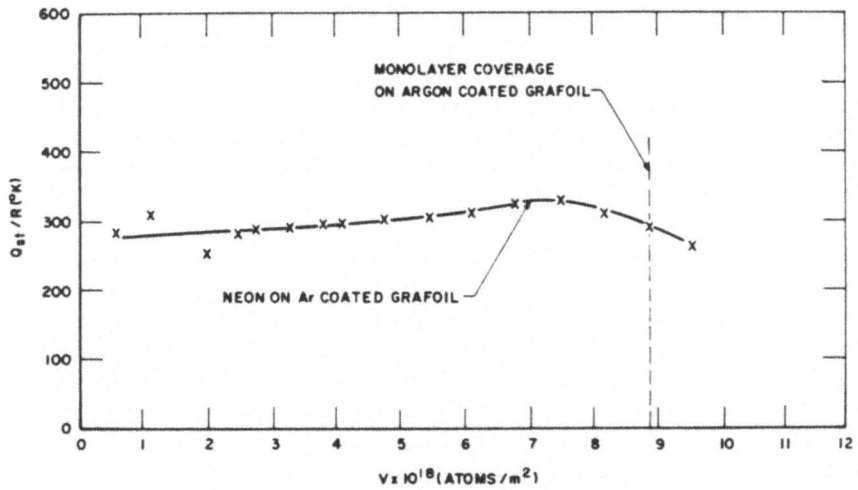

Fig. 3. Isosteric Heat of Adsorption of Neon on argon-
coated Grafoil as a function of coverage.

From these figures and other data obtained by us extrapolations have been of (Q_{st}/R) as the coverage goes to
zero (i.e. for p→o) and the values so obtained of
$(Q_{st}/R)_{p→o}$ are tabulated in Table I along with the
values, where known, of the monolayer coverages.

TABLE I

Adsorbate and Substrate	$(Q_{st}/R)^*_{p→o}$ (°K)	E_b (Theor) (°K)	Monolayer Coverage (atoms/m^2)
^{4}He on Copper	∿160	–	10.2 x 10^{18}
^{4}He on Ar-coated Copper	76	62	7.9 x 10^{18}
^{4}He on Grafoil	170	189	11.2 x 10^{18}
^{4}He on Ar-coated Grafoil	82	71	8.3 x 10^{18}
Neon on Copper	395	–	9.9 x 10^{18}
Neon on Ar-coated Copper	320	–	8.3 x 10^{18}
Neon on Grafoil	–	–	–
Neon on Ar-coated Grafoil	280	–	8.9 x 10^{18}

*At $T_{AV} ≈ 10°K$.

Figs. 1, 2 and 3, together with Table I, form a convenient summary of such data. The table moreover includes
the results of some theoretical calculations of the
binding energy, E_b in the limit of zero coverage. It
is to be noted parenthetically here that E_b is the
same as $(Q_{st}/R)_{p→o}$ when T→o. Some discussion of the
individual Figs. follows.

First to compare with data obtained elsewhere,
where possible, we note first that E_b at about $T_{AV} ≃ 10K$
has been experimentally determined by Pollock and co-
workers[5] by a quite different method and their result
is in substantial agreement with that deducable from
our data. Secondly Elgin and Goodstein[13] has obtained
(Q_{st}/R) data for ^{4}He on Grafoil, again in substantial
agreement with our results. No other detailed data
are available, particularly for the argon-coated
substrates.

One striking general conclusion of our results is that the argon-coated substrates, whether copper or Grafoil, show greater uniformity and homogenity in adsorption properties over a greater range of coverage than the corresponding bare substrates. This is true for both ^{4}He and neon as adsorbates and is evident by the flatness of the (Q_{st}/R) versus coverage plots over a wide range of coverage. Because of this outstanding result, it is considered that future work would do well to investigate other properties of adsorption on argon-coated substrates, particularly the argon-coated Grafoil.

Another interesting result, evident in Table I is that the values of $(Q_{st}/R)_{p\to o}$ for ^{4}He on the argon-coated substrates (i.e. argon-coated copper and argon-coated Grafoil) are nearly the same. It is evident that the argon, although only a monolayer, is dominant in determining the adsorption energy.

A further associated result is that the monolayer coverages of ^{4}He on the two argon-coated substrates are very similar in numerical value. Moreover, the values of the monolayer coverages of neon on the two argon-coated substrates are close to those of ^{4}He.

The data for E_b obtained theoretically and given in Table I seems to fit well with the $(Q_{st}/R)_{p\to o}$ data when allowance is made for the finite average temperature (10K) of the (Q_{st}/R) data for ^{4}He on the argon-coated substrates. For ^{4}He on the bare Grafoil there seems to be a discrepancy between theory and experiment.

PHASE TRANSITIONS DETERMINED BY ADSORPTION ISOTHERMS

Observations of phase transitions through measurement of adsorption isotherms have been made over the past many years, particularly in the temperature range of about 65K to 120K. In general phase transitions, observed in this way, have been confined to adsorption on very homogeneous substrates, as exemplified by small crystals (about 5 microns on edge) of alkali halides precipitated from solution, by graphite and by exfoliated graphite (Grafoil.) Some of the major examples of these observations are reviewed here to provide some background illumination ot the problem of the phases of submonolayer helium adsorbed on such homogeneous substrates.

Some of the earliest observations of phase transitions in submonolayer adsorbed gases on graphite by adsorption isotherm techniques were those of Jura and Criddle[14] in 1951 who examined argon on graphite and of Ross and Winkler[15] in 1955 who studied krypton on graphitized carbon black. Fig. 4 shows some of the results of the former authors which plots coverage versus vapor pressure for submonolayer argon on graphite at temperatures, as marked, in the range 63.8K to 77.3K.

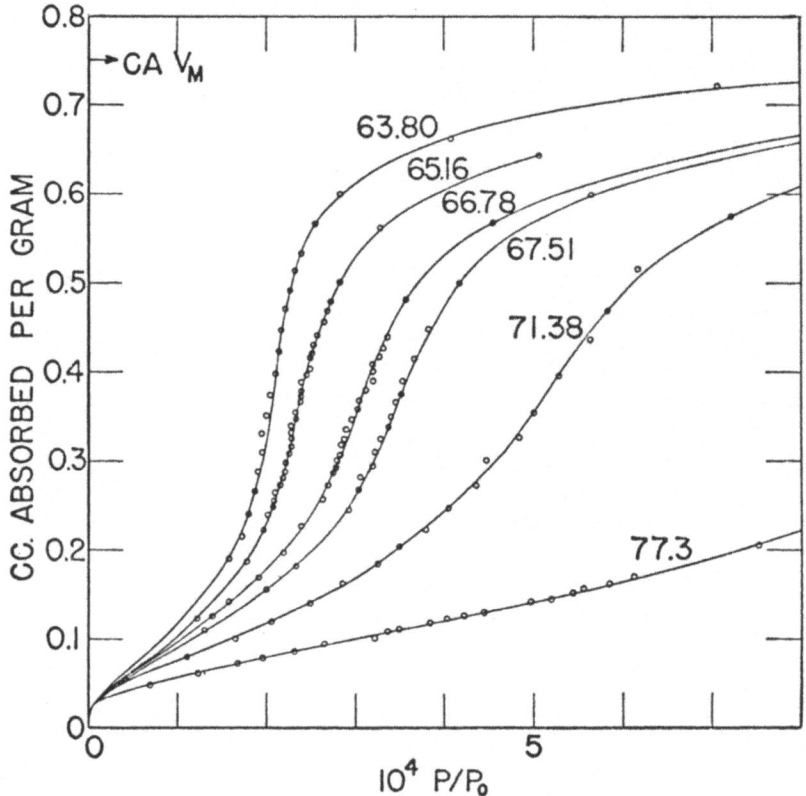

Fig. 4 Adsorption isotherms of argon on graphite at the temperatures marked (Kelvin degrees) according to Jura and Criddle[14]

Note that the monolayer coverage is 0.75ml/g. The similarity of this family of curves to what one might expect for a two-dimensional van der Waals gas is striking and the curves in Fig. 4 with a vertical, or near-vertical, section are interpreted as follows: starting from the origin, the portion of the curve up to the region where it starts to turn vertically upwards is

taken to characterize a mobile gas-like phase of the adsorbate; where the curve starts vertically upwards is
taken to represent the beginning of a condensation in
phase of the adsorbate, which condensation proceeds to
completion as the curve is followed vertically. This
vertical section therefore represents a two-phase region. At the location where the upper vertical part
of the curve bends away to the right, the condensation
is assumed to be complete and the curve thereafter again
reflects a single (condensed) phase. Similarly shaped
adsorption isotherms have been observed by Ross and
Winkler[15] for krypton on graphitized carbon black in
the temperature range 77K to 90K, and they interpreted
their results in a similar manner. They deduced the
critical condensation temperature, T_c, for this system
was 82K.

These authors supposed that the condensed phase was
in the form of a two-dimensional lattice in register
with the exposed plane of the graphite substrate lattice,
and that those parts of the isotherms to the right of
the vertical two-phase sections of the curves reflected
a process whereby a shifting of the atoms forming the
overlaying lattice occurred in order to permit occupancy
of the last portions of the adsorption surface, i.e.
a breakdown of the adsorbate lattice occurred to allow
filling of the first layer to monolayer completion.
The occurrence of such adsorbate lattices is confirmed by
the elegant specific heat measurements by Dash and co-
workers[17, 18] on similar systems--namely, submonolayer
helium and neon on Grafoil. Such lattice transitions
may be an alternate interpretation of the vertical
changes occurring in many adsorption isotherms of the
rare gases on exfoliated graphite, as observed in 1970
by Thomy and Duval[19].

To continue, many adsorption isotherms of the kind
shown in Fig. 4, revealing the occurrence of phase
transitions, have been observed on the very homogeneous
substrates produced by precipitation crystallization of
alkali halides from solution. These phase transitions
were observed in the temperature range 65K to 120K and
were confined to adsorbates which were exactly or almost spherically symmetrical atoms or molecules, as
for example argon, krypton, xenon, ethane and methane.
Unsymmetrical molecules, as for example nitrogen, did
not show such transitions under identical ambient conditions. Some of the major observations, revealing
phase transitions in the temperature range cited,
(where available the critical condensation temperature,

T_c, is also quoted), were reported for systems as follows:
In 1954, argon on GaF₂ for 79K<T<112K by Edelhoch and
Taylor[20]; in 1954, xenon, ethane and methane on NaCl
for 78K<T<148K by Ross and Clark[21] (T_c values as follows: xenon 104K, ethane 132K and methane 90K); in 1954,
ethane on NaCl at 90 K by Ross and Winkler[22]; in 1957
and 1958, krypton on NaBr for 67K<T<86K by Fisher and
McMillan[23,24] ($T_c \simeq$ 80K); in 1967, krypton on NaCl,
on RbCl and on KCl at 78K by Takaishi and Saito[25].

Of the above investigations, that reported in most
detail both experimental and theoretical is the work of
Fisher and McMillan[24]. Their adsorption isotherm data
for submonolayer krypton on NaBr is shown in Fig. 5,
which shows clearly the characteristics of condensation
described previsously for submonolayer krypton on
graphite (Fig. 4).

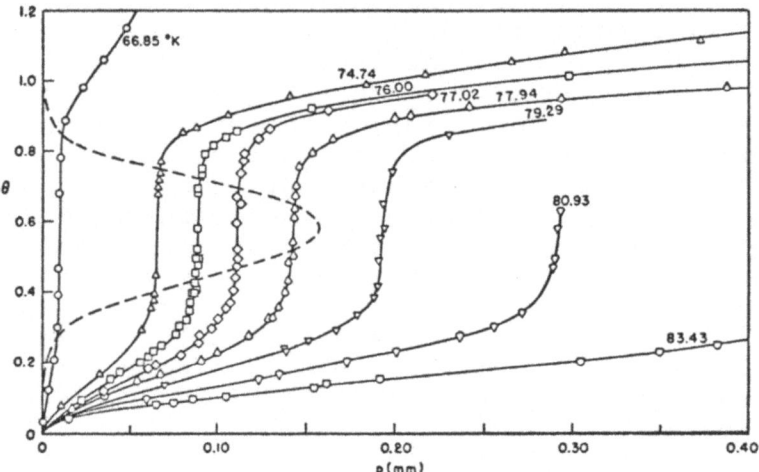

Fig. 5. Adsorption isotherms of krypton on NaBr at
the temperatures marked according to Fisher and McMillan[24].

The broken curve of Fig. 5 gives Fisher & McMillan's
computation of the two-phase region. By comparing
these results with theoretical models, the authors were
able to calculate the two-dimensional krypton-krypton
interaction forces.

Now coming to more recent work and switching back
to graphite substrates, the following adsorption isotherm investigations extended previous work in range
and in detail. First there is the extensive adsorption

isotherm work on exfoliated graphite by Thomy and Duval[19, 26-28] published between 1964 and 1970. They observed phase changes for krypton, xenon and methane submonolayers in the temperature range 77K to 118K. One of their data plots is shown in Fig. 6 for xenon on exfoliated graphite.

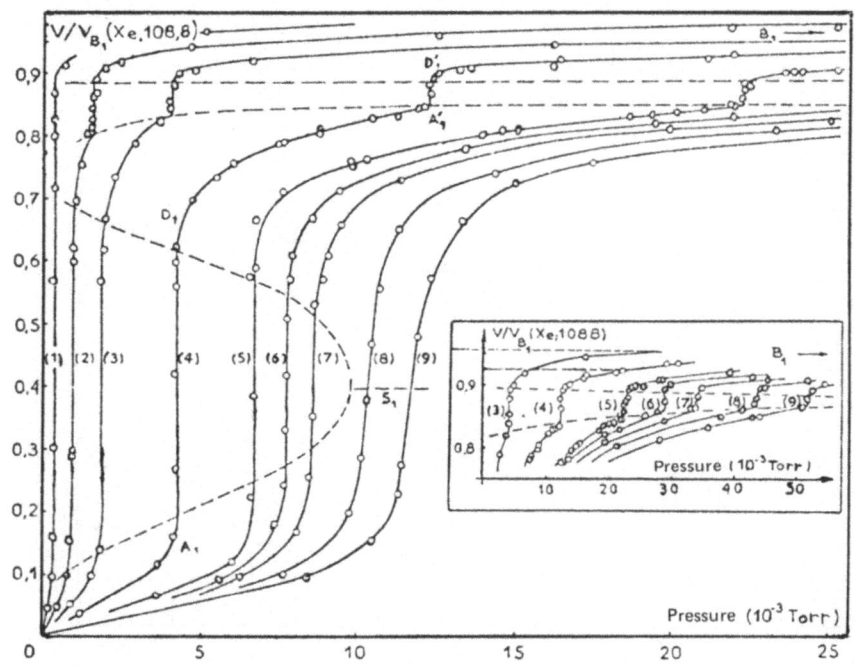

Fig. 6. Adsorption isotherms of xenon on exfoliated graphite by Thomy and Duval[19] at temperatures as follows: (1) 77.3K; (2) 82.4K; (3) 84.1K; (4) 85.7K; (5) 86.5K; (6) 87.1K; (7) 88.3K; (8) 89.0K; (9) 90.1K; (10) 90.9K.

The main features of Figs. 4 and 5 are again reproduced here. In addition, on several of the isotherms there is a secondary vertical section of the curves, to the right of the primary vertical sections, which secondary section Thomy & Duval interpreted as due to a liquid solid phase transition, having assumed that the primary vertical section represents a gas to liquid transition.

A brief note, giving a single adsorption isotherm for submonolayer neon on exfoliated graphite was authored by Thomy, Duval and Regnier[28] in 1969 which showed a

small step, which was interpreted by them to be indicative of a phase transition. This has recently been followed by a report on adsorption isotherm data for submonolayer neon on Grafoil coated with an argon monolayer
by Lerner, Hegde and Daunt[4]. Fig. 7 shows this data.

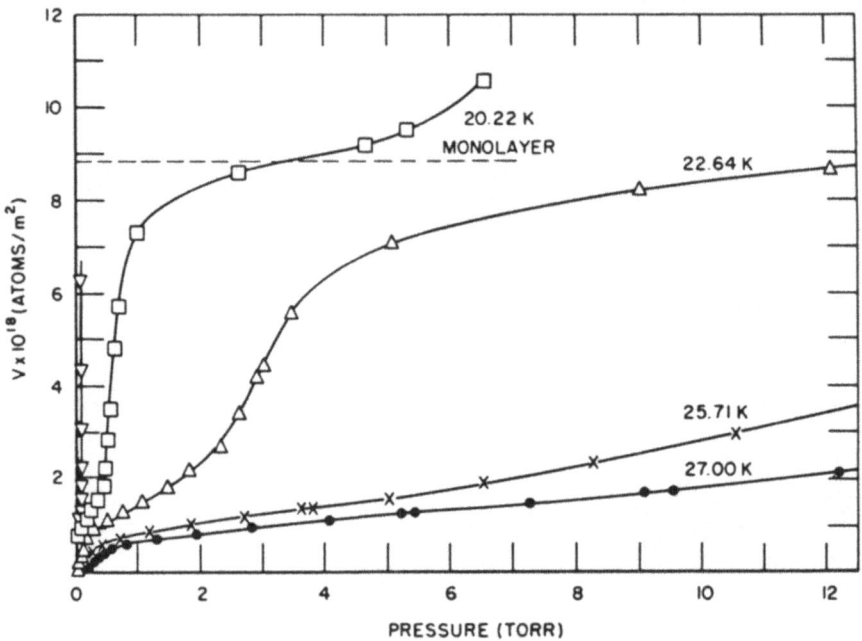

Fig. 7. Adsorption isotherms of neon on argon-coated
Grafoil at temperatures as marked[4].

It seems apparent that this indicates a phase transition,
with a critical condensation temperature, T_c, of 21.5K
The existence of a phase transition in monolayer neon
on bare graphite has been shown by specific heat measurements , notably by Steele and Karl[29], by Antoniou
et al[30], and by Huff and Dash[18]. The detailed correlation of the transition mechanisms, as revealed by the
two types of measurements, remain to be worked out.

      Going to lower temperatures, adsorption isotherm
measurements of submonolayer ^{4}He on Grafoil coated with
an argon monolayer by Lerner and Daunt[7] down to 4.2K
are shown in Fig. 8. Although these data do not present the same signature as given in previous figures
for a phase transition, there remains the possibility

that an extension of this work to lower pressures would
reveal a transition.

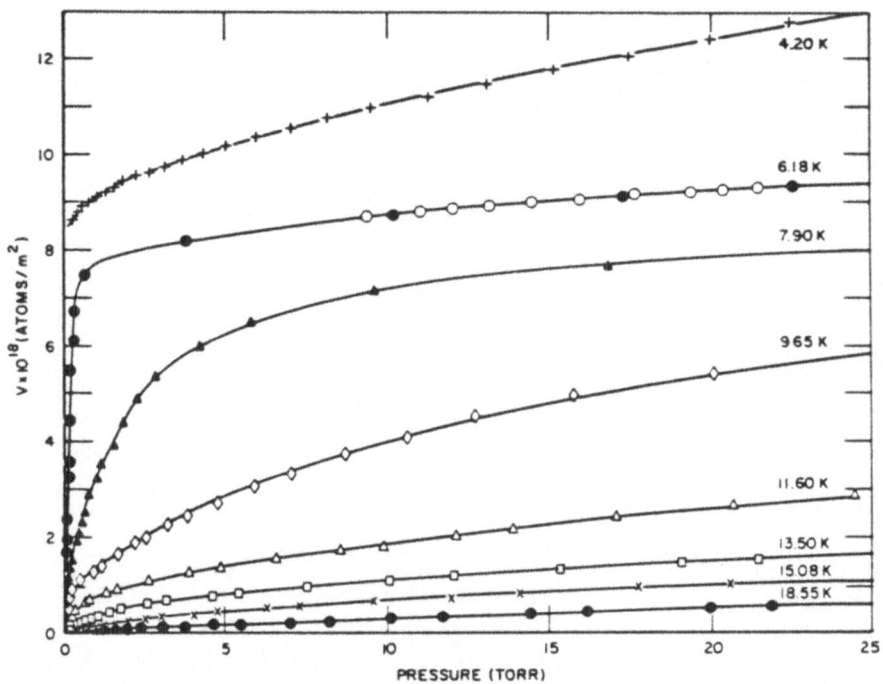

Fig. 8 Adsorption isotherms of ⁴He on argon-coated
Grafoil at temperatures as worked as reported by Lerner
and Daunt[7].

TRANSPORT PROPERTIES OF GRAFOIL

In view of the fact that the Grafoil substrate is
used a great deal at low temperatures, it was thought
that some measurements of its thermal and electrical
conductivities at low temperatures would be of value in
order to assess quantitatively such questions as thermal diffusion times in calorimeter and rf penetration
of NMR signals. Our data[10] are given in Figs. 9 and
10, which give the longitudinal electrical conductivity,
σ_L, and the longitudinal κ_L, and transverse, κ_T, thermal
conductivities, as measured on a 0.005 inch thick

Grafoil sample in the temperature range 4K to 300K.

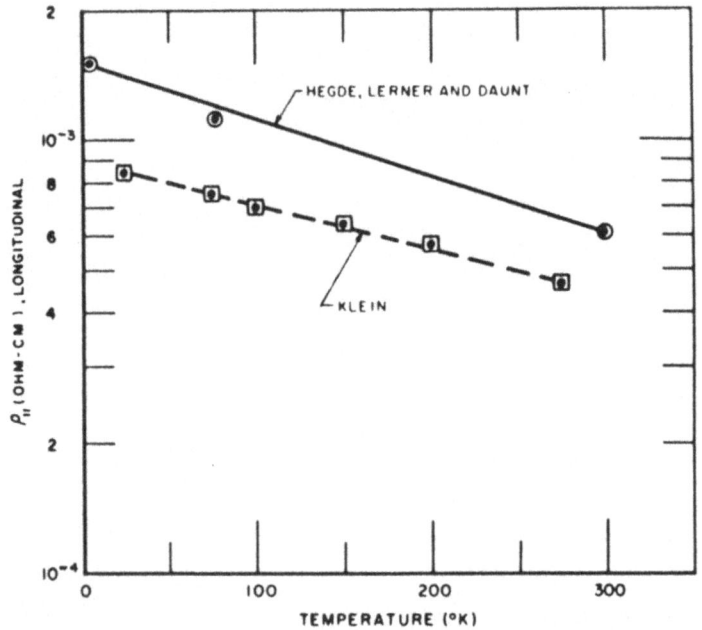

Fig. 9 The longitudinal electrical resistivity of Grafoil[10). The curve labeled Klein is for pyrolytic
graphite[31).

Fig. 9 gives for comparison the only previous comparable
data, that for σ_L, by Klein[31) on a sample of "as-deposit-
ed pyrolytic graphite". Our data can be approximately
described by the following equations:

$$\sigma_L = 640.3 \exp(0.003T) \quad \text{mho/cm}$$
$$\sigma_T = 1.84 \exp(0.003T) \quad \text{mho/cm}$$
$$\kappa_L = 0.075 \times 10^{-3} T^{0.94} \quad \text{W/cm-K}$$
$$\kappa_T = 0.367 \times 10^{-6} T^{1.8} \quad \text{W/cm-K}$$

At 4.2K the Lorenz number, $L = \kappa\sigma/T$, is 10.5×10^{-8}
watts ohm/deg.^{2} for the longitudinal measurements and
71×10^{-8} for the transverse. These are to be compared
with a theoretical value for an ideal system with only
electronic conductivity of 2.45×10^{-8} watts ohms/deg.^{2}.
These results indicate that the lattice thermal conductivity in Grafoil is dominant.

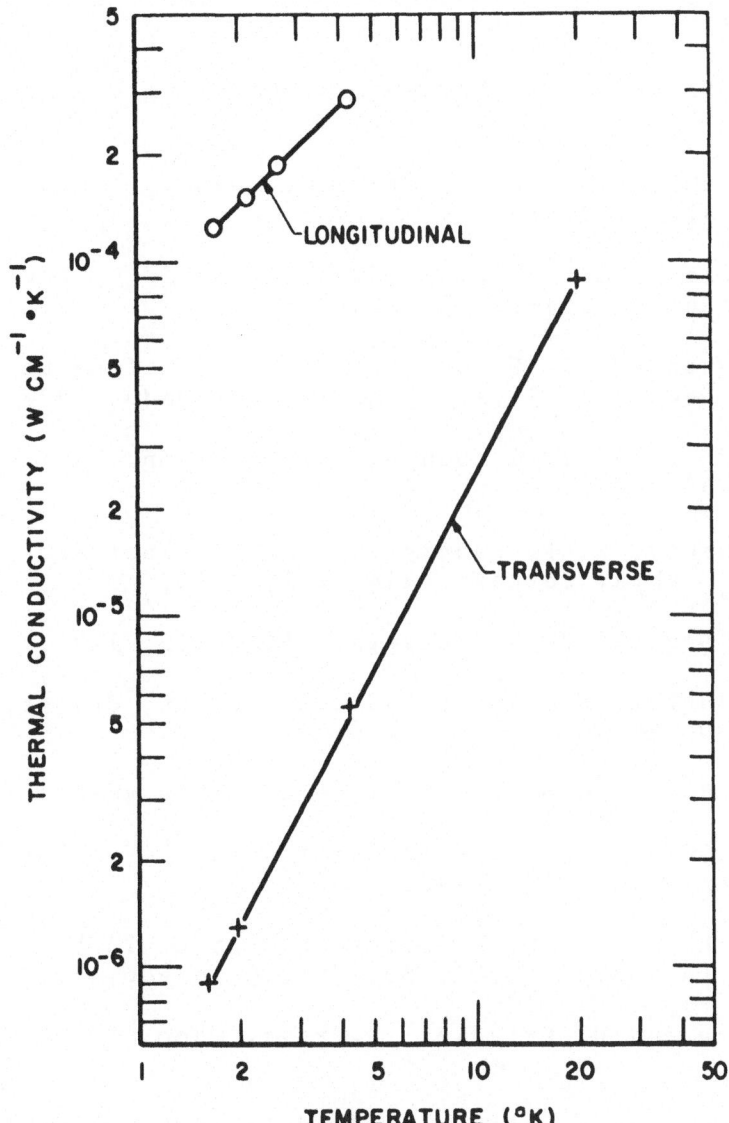

Fig. 10. The longitudinal and transverse thermal conductivities of Grafoil [10].

REFERENCES

1) J. G. Daunt and E. Lerner. "Adsorption Desorption Phenomena" Publ. Academic Press, 1972, p. 127.

2) E. Lerner and J. G. Daunt. J. Low Temp. Phys. 6, 241, 1972.

3) J. G. Daunt and E. Lerner. J. Low Temp. Phys. 8, 79, 1972.

4) E. Lerner, S. G. Hegde and J. G. Daunt. Phys. Letters. 41A, 239, 1972.

5) F. Pollock, H. Logan, J. Hobgood and J. G. Daunt. Phys. Rev. Letters. 28, 346, 1972.

6) E. Lerner and J. G. Daunt. Proc. 13th Internat. Conf. Low Temp. Phys., Boulder, Colorado, 1972.

7) E. Lerner and J. G. Daunt. J. Low Temp. Phys. 10, 299, 1973.

8) P. Mahadev, M. F. Panczyk, R. A. Scribner and J. G. Daunt. Phys. Letters. 41A, 221, 1972.

9) J. G. Daunt. Phys. Letters. 41A, 223, 1972.

10) S. G. Hegde, E. Lerner and J. G. Daunt, Cryogenics 13, 230, 1973.

11) D. L. Husa and D. C. Hickernell. This volume, p 133.

12) D. L. Goodstein, W. D. McCormick and J. G. Dash, Cryogenics, 6, 167, 1966.

13) R. L. Elgin and D. L Goodstein. Proc. 13th Internat. Conf. Low Temp. Phys., Boulder, Colorado, 1972.

14) G. Jura and D. Criddle. J. Phys. Chem. 55, 163, 1951.

15) S. Ross and W. Winkler. J. Colloid Science. 10, 330, 1955.

16) See, for example, J. G. deBoer. "The Dynamical Character of Adsorption." Oxford Press, 1968.

17) M. Bretz and J. G. Dash. Phys. Rev. Letters. 27,
     647, 1971.

18) G. B. Huff and J. G. Dash. (Preprint)

19) A. Thomy and X. Duval. J. Chim. Physique. 67,
     1101, 1970.

20) H. Edelhoch and H. S. Taylor. J. Chem. Phys. 58,
     344, 1954.

21) S. Ross and H. Clark. J. Am. Chem. Soc. 76, 4291,
     1956.

22) S. Ross and W. Winkler. J. Am. Chem. Soc. 76,
     2637, 1954. (See also reference 21)

23) B. B. Fisher and W. G. McMillan. J. Am. Chem. Soc.
     79, 2969, 1957.

24) B. B. Fisher and W. G. McMillan. J. Chem. Phys.
     28, 549, 555, and 562, 1958.

25) T. Takaishi and M. Saito, J. Phys. Chem. 71, 453,
     1967.

26) A. Thomy and X. Duval. J. Chim. Physique. 66,
     1966, 1969.

27) X. Duval and A. Thomy. Comp. Rendues. 259, 4007,
     1964.

28) A. Thomy, X. Duval and J. Regnier. Comp. Rendues.
     268, 1416, 1969.

29) W. A. Steele and R. Karl. J. Colloid & Interface
     Sci. 28, 397, 1968.

30) A. A. Antoniou, P. D. Scaife and J. M. Peacock,
     J. Chem. Phys. 54, 5403, 1971.

31) C. A. Klein. J. Appl. Phys. 35, 2947, 1964.

^{4}HE ON GRAFOIL: THE MELTING TRANSITION *

Robert L. Elgin and David L. Goodstein

Physics Department, California Institute of Technology

Pasadena, California 91109

Recent measurements of both the heat capacity and vapor pressure of helium adsorbed on Grafoil have been of sufficient accuracy, reproducibility, and completeness to allow characterization of all the static thermodynamic properties of the two-dimensional film above 2 K. Interactions between the film and the bulk gas have been accurately modeled, showing that the first layer of the film behaves even more like the bulk solid than was previously apparent. The only major difference is that the transition from crystal to fluid is no longer of first order. Two recent theories of melting are discussed. It is suggested that the growth of dislocations may be the cause of melting in both two and three dimensions and the true basis for the empirical Lindemann formula.

1. INTRODUCTION

The large improvement in understanding of submonolayer films occasioned by the use of exfoliated graphite has been well documented. As recently as 1969, there was still considerable doubt whether phase transitions could occur in films adsorbed on solids. (1) At that time, Thomy and Duval were in the midst of pressure isotherm experiments using argon, krypton, and xenon that showed phases with critical points and triple points, in remarkable analogy to bulk behavior. (2) More recent work by Bretz, Huff, and Dash, using heat capacity measurements, has shown that for ^{4}He there is a transition, not only analogous to, but practically coincident with, the melting line in the bulk. (3) We hope to show that these experiments have opened a rich new field and that the thermodynamic properties of genuinely two-dimensional matter may be determined from such experimental data.

2. TECHNIQUES

The problem of deducing the properties of an idealized two-dimensional film from experimental data may be separated into three parts: direct measurement of the properties of the entire adsorption cell, isolation of the contributions arising specifically from the film, and identification of that part of the film contributions arising from interactions between the film and the substrate potential. The first part is purely an experimental problem. Complete thermodynamic information may be obtained by combining heat capacity and vapor pressure measurements, provided they overlap in temperature and coverage (two-dimensional density). The second part may be made completely model-independent by use of the formalism of surface excesses. (1) The third part requires the introduction of theories about the interactions. However, if the data are complete enough, one can usually find regions where the measurements are sensitive to only one aspect of the interactions at a time. This allows the complete correction for interactions to be built up with confidence.

The experimental apparatus consists of a pile of Grafoil (4) sheets in a copper cell suspended in a low-temperature vacuum enclosure. Provision is made to introduce measured amounts of helium gas and heat, and to measure the pressure and to measure and regulate the temperature. A detailed description and all the raw data appear elsewhere. (5)

The thermodynamic function measured is the Landau potential,

$$\Omega = F - \mu N \ ,$$

where F is the free energy, N is the number of particles, and μ is the chemical potential. For rigid walls and substrate, its differential equation is (6)

$$d\Omega = - S \ dT - N \ d\mu \ .$$

At high temperatures, μ is deduced from vapor pressure data and Ω is found by integration along isotherms from very low coverage. The entropy may be computed by comparing two adjacent isotherms, and extended to lower temperatures by use of the heat capacity measurements. One may then use the Maxwell relation,

$$(\partial S/\partial N)_T = - (\partial \mu/\partial T)_N \ ,$$

to find dμ and hence the Landau potential at all temperatures. The free volume, V, may be measured by gas displacement at high temperatures, where the adsorption is negligible. The gas contributions to N, S, and Ω may be calculated from T, V, and the pressure, P and the experimental three-dimensional equation of state. (7) (The gas contribution to Ω is simply - PV.) Subtracting these from the corresponding measured quantities gives the surface excess values.

Three of the 14 pressure isotherms constructed in this manner are shown in Fig. 1. The coverage has been put on a molecular area scale by use of the sharp heat capacity peak that occurs at 2.92 K when the coverage is precisely one helium atom for every 6 surface carbon atoms (the lattice gas ordering transition). The helium-helium attractive energy is so small relative to the zero-point energy and the binding energy that the melting transition can just barely be seen on two of the curves. It would clearly be helpful if we could first evaluate these large energy terms. This is simplest at absolute zero, where the heat capacity peak disappears. Unfortunately, it has not yet been possible to build a single apparatus that can measure heat capacity accurately from above 10 K down to the mK range. Instead, high temperature data obtained by the authors have been combined with data obtained at the University of Washington using adiabatic demagnetization (8) and a helium dilution refrigerator. (9)

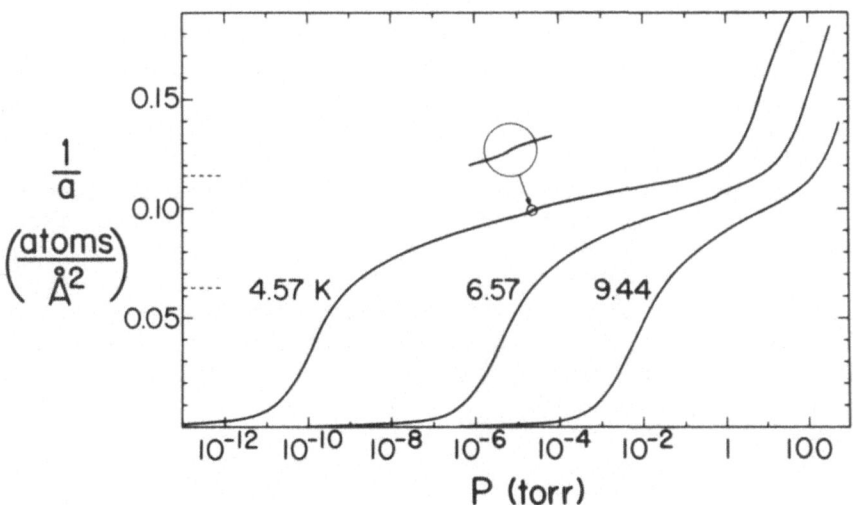

FIG. 1. Typical pressure isotherms. The data above 0.001 torr were measured directly. The lower pressures were found using the heat capacity measurements as explained in the text. All experimental points fall within the widths of the lines shown. The circle shows the melting transition at 4.57 K. The transition occurs near 1 torr at 6.57 K. The dashed lines show the coverages of lattice gas ordering and monolayer completion.

With Grafoil (and presumably with other exfoliated graphite)
it is possible for the first time to characterize separate substrate samples to an accuracy such that derivatives of the thermodynamic functions are not seriously degraded by the use of combined
data. There are several reasons for this. Over 90% of the substrate appears to consist of perfect basal-plane graphite, so it
can be described by a single parameter, the area. The relative
areas of two different substrates can be measured to a fraction of
one per cent by means of the lattice gas ordering transition, which
occurs at a convenient intermediate temperature. The remaining
inhomogeneity appears to be represented well by a single parameter,
the decrease in area in the second layer due to crevices. (10)

The chemical potential of the helium film at zero kelvin which
results from this combined analysis is shown in Fig. 2. The chemical potential of bulk helium is also shown. It may be found from

$$\mu = - L_0 + \int_0 v \, dP \, ,$$

using low temperature data (11), where L_0 is the latent heat at
zero kelvin, and v is the molar volume. The similarity to Fig. 1
is not accidental, since μ is proportional to log P. From the bulk
curve, one can extract the energy and compressibility of the liquid
and solid as a function of the density. (There is no gas phase at
absolute zero.) One can extract similar information from the film
curve, and much more. The compressibility at any coverage is given
by

$$K_2 = - \left(\partial a / \partial \mu \right)_T \, .$$

A local minimum in the compressibility occurs at the lattice gas
ordering coverage. According to Novaco, the helium-helium interaction energy is less than 1 K at all lower coverages. (12) Therefore μ is equal in magnitude to the binding energy in this region.
For most of the first layer, this means the binding energy is about
144 K. However, a small percentage of the sites still show large
inhomogeneity, as the binding energy rises as high as 200 K. Similarly, at the top of the figure, we see that the binding energy of
the dilute second layer is about 29 K.

3. RESULTS AND ANALYSIS

The experimental heat capacity of the dense film as reported
by Bretz, Huff, and Dash, (3) uncorrected for desorption, is shown
in Fig. 3. A confusing range of behavior is seen. At low temperatures, the heat capacity is generally proportional to T^2. At intermediate temperatures, there is a progression of small, broad
peaks at increasing coverage. At high temperatures, the peaks are
large and narrow and the heat capacity is always far above the

ideal gas value. We will now show that we can subtract out the
effects of interactions with an ideal two-dimensional gas in the
second layer and an ideal three-dimensional gas in the free volume,
leaving a much simpler heat capacity that scales with temperature
and density as it does in the bulk case.

If the experimental points in Fig. 3 were as closely spaced in
coverage as they are in temperature, the following procedure could
be used. Find the Landau potential of the film directly by numerical integration and differentiation as described in section 2.
Assume the second layer to be an ideal two-dimensional gas with the
same total area as the first layer and with the binding energy
given by Fig. 2. Its thermodynamic properties are then completely
specified, so its contribution to N, S, and Ω may be calculated
from μ and T using statistical mechanics. Subtract these contributions from the total film values to leave a description of the
first layer. Then find the heat capacity by differentiating S.

Lacking such experimental data, an interpolation procedure had
to be devised. The small heat capacity peaks at intermediate temperature vary systematically, so it was not difficult to construct
a formula that reproduced them. Applying the procedure given above

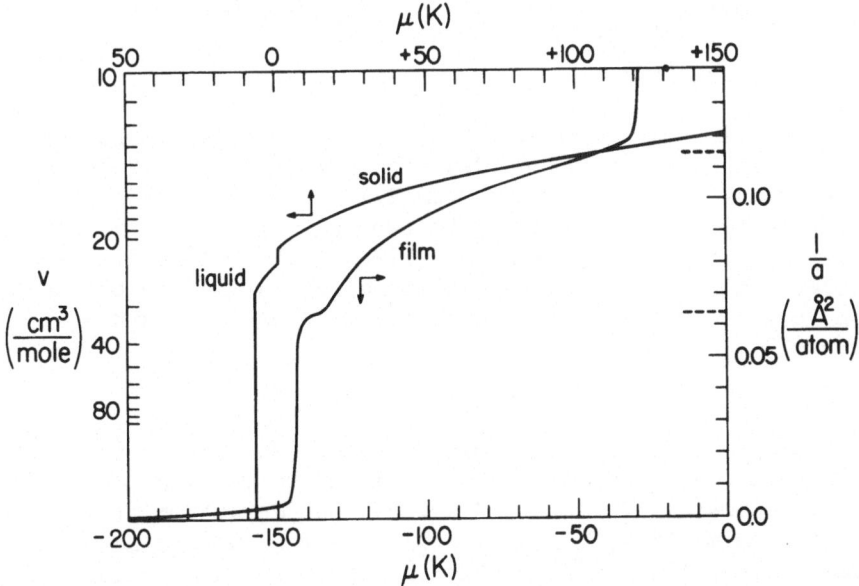

FIG. 2 . The chemical potentials of bulk helium and the film at
absolute zero. The bulk molar volume scale has been adjusted to
correspond to the same interatomic spacing as the film coverage.
Note the zero offset on the horizontal scales.

showed that both the second layer and the bulk gas corrections were negligible in this region. It was decided at this point to test the hypothesis that the first layer was very simple. The formulae that fit the intermediate temperature peaks for the experimental system were simply assumed to describe the first layer at all densities. The procedure in the preceding paragraph was reversed and heat capacity curves for the film-vapor system were generated. The peaks became much larger above 6 K just as they do experimentally. By small adjustments in the extrapolations and in the free volume, it was found possible to match all the experimental peak temperatures. (It was later found that this best fit free volume agrees within 6% with the actual value measured in Seattle. (13)) The resulting calculated curves are shown in Fig. 3. The details of the model follow.

From the film curve in Fig. 2, we find the entropy, chemical potential, and internal energy at absolute zero to be given by

$$S(0) = 0 ,$$

$$\mu(0) = -141 + 43.6 \, N^5 ,$$

and $\qquad U(0) = -3.5 - 141 \, N + 7.27 \, N^6$, for $0.9 < N < 1.15$,

where Boltzman's constant has been set equal to 1 and N is the number of atoms of helium per 10.3 $\mathring{A}^2$ of surface. Therefore $1/a$ in atoms/$\mathring{A}^2$ is given by $N/10.3$. From the Debye temperatures listed by Bretz, Huff, and Dash (3)

$$\theta_D = 33 \, N^{3.5} , \quad \text{for} \quad 0.9 < N < 1.18 .$$

The exponent in this formula should be the Gruneisen constant. For bulk helium in the equivalent density range (13.5 to 18 cc/mole), this "constant" rises smoothly from 2.12 to 2.56 . (14) As a function of the interatomic spacing in two and three dimensions respectively, the 3.5 becomes 7 and the 2.12 to 2.56 become 6.36 to 7.68. So we see that the two and three dimensional Gruneisen constants are in very close agreement. From the melting temperatures in reference 3,

$$T_m = 4.35 \, N^5 , \quad \text{for} \quad 0.9 < N < 1.05 .$$

This is also consistent with bulk behavior (see Fig. 5 in section 4). The entropy change of the first layer on melting can be defined approximately in terms of the maximum deviation from the entropy of a nonmelting, harmonic solid. For a two-dimensional triangular close-packed lattice, the entropy of such a solid is uniquely defined in terms of the zero kelvin Debye temperature. (15) However, the resulting function cannot be integrated analytically. For purposes of this analysis, the heat capacity of the solid is taken to be $3 \, N^4/(115 \, N^{10}/T^2 + 1)$ and the entropy of melting to be $0.46 - 0.96/T_m$. Aside from a slight rounding at the

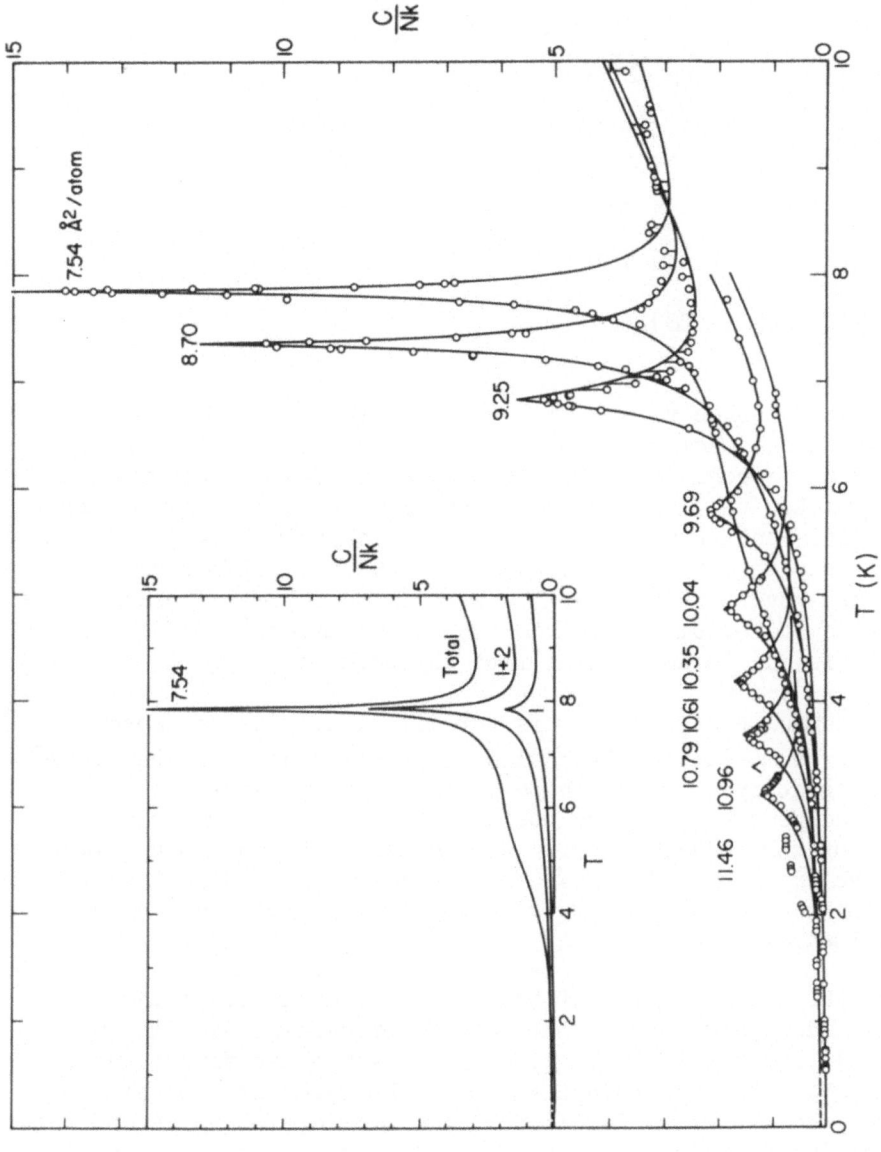

FIG. 3. The heat capacity of the dense film. The data points are from reference 3. The solid lines were calculated as explained in the text. The inset shows the contributions of the first and second layer to the highest peak.

maximum, the small heat capacity peaks at intermediate temperature fit well to an exponential form. The overall result is:

$$C = 3 \ N^4/(115 \ N^{10}/T^2 + 1) + (0.46 - 0.96/T_m)(T/T_m)M$$

where $M = 5.40 \ \exp(9.5(T/T_m - 1))$ for $T < T_m$

and $M = 5.40 \ \exp(12.5(1 - T/T_m))$ for $T > T_m$.

One may then find μ from:

$$U = U(0) + \int C \ dT$$

$$S = S(0) + \int C \ d(\log T)$$

and $\mu = (\partial(U - TS)/\partial N)_T$.

Setting the free volume at 10 cm^3 per 277 m^2 of surface area and requiring the chemical potential of the ideal gases in the free volume and the second layer to match that given above for the first layer, completes the model. The contributions to the heat capacity from the different terms are shown by the inset in Fig. 3. Here we see explicitly that the heat capacity in the first layer remains small at all coverages (curve 1). Part of the observed peak is due to promotion of helium atoms into the second layer (curve 1+2), but most of it is due to desorption into the bulk gas.

The assumed heat capacity is unrealistically large above 2 T_m, but at lower temperatures it scales as promised and its predictive ability is gratifying. The shape of the large peaks was not adjustible independently of the position of their centers and the the shape of the smaller peaks. Nonetheless, the curves go right through the experimental data. The shoulder near 6 K in the inset was totally unexpected, but it also is in complete agreement with the experimental points.

From the heat capacity of the first layer alone, we may calculate the Landau potential by the methods of section 2. The two-dimensional pressure is then given by $\phi = - \Omega/A$, in direct analogy to $P = - \Omega/V$. The results are shown in the right half of Fig. 4. The corresponding three-dimensional isochores are shown on the left half. One immediately sees that the film expands when melting at constant pressure, just as in the bulk case. At the lower coverages, the adsorption forces hold the film at constant density, so the pressure rises instead. However, at higher coverages the process of melting overcomes the adsorption forces enough to promote some helium atoms to the second layer. This does work against the adsorption forces and also reduces the first layer density, which

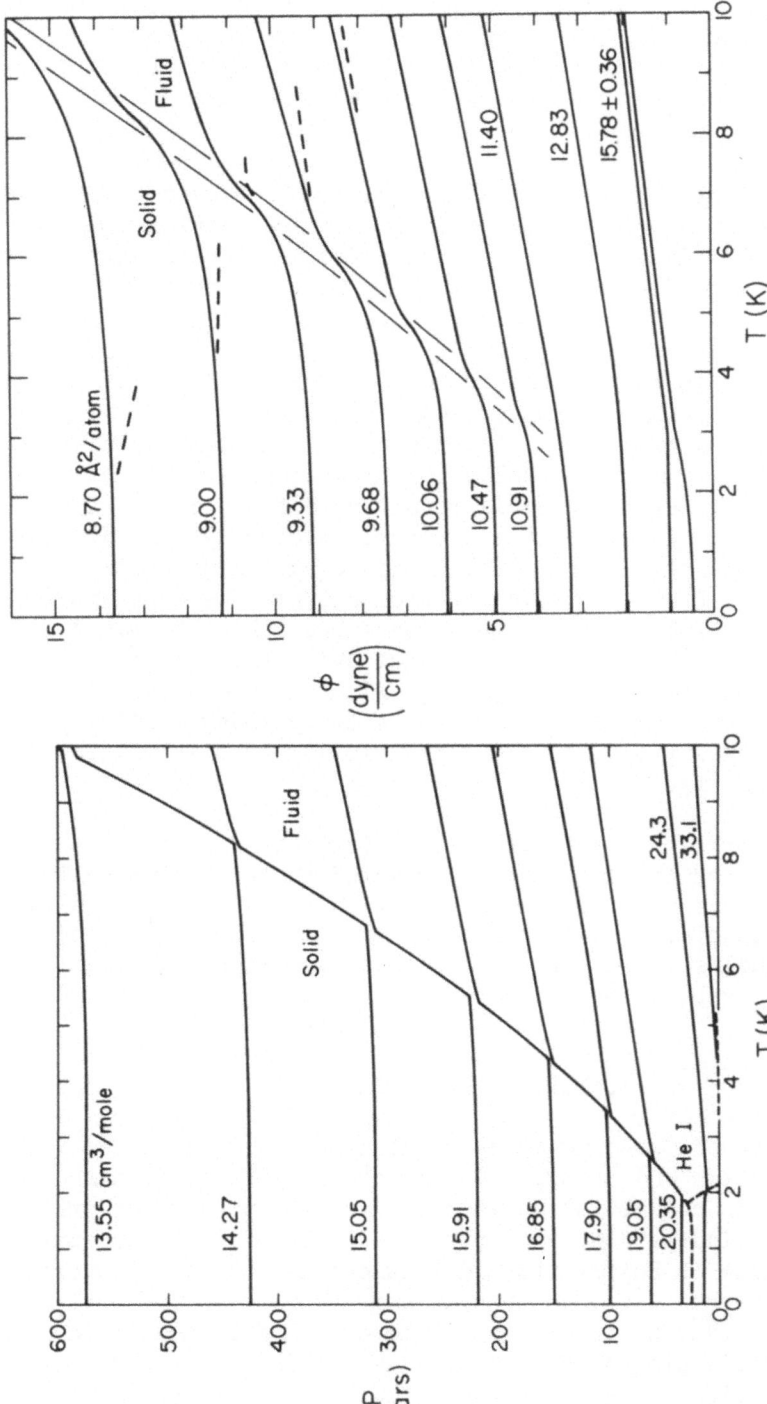

FIG. 4. Three and two dimensional isochores at high density. The curves in the two plots are drawn at corresponding interatomic spacings. The dashed lines in the upper right of the two-dimensional plot are uncorrected for multilayer formation. The solid lines have been extended to higher temperatures by use of the model in this section. The lower coverage curves have been calculated directly from experimental data using the methods of section 2. The bottom pair of curves bracket the lattice gas ordering transition.

in turn lowers the melting temperature. These effects combine to increase the apparent height and reduce the width of the heat capacity peaks. In addition, the presence of the second layer causes the melting peaks to appear to deviate from the bulk melting curve. When T_m is plotted versus the density in the first layer alone, this deviation disappears.

Helium adsorbed on Grafoil at near-monolayer coverages was originally reported to have Debye temperatures and melting temperatures similar to bulk helium. (3) More recently, the two-dimensional elastic constants have been shown to be consistent with those of the bulk (16) and the entropy change on melting has been found to be about one-third as large as in the bulk. (17) The Gruneisen constants have also been found to correspond. (18) In this paper we have filled out this comparison. Along with the Gruneisen constant, the compressibility has nearly the same value in two and three dimensions when scaled for dimensionality. (This can be deduced from Fig. 2.) The small remaining differences in the Debye temperatures may be due to a geometric factor between hcp and triangular close-packing (tcp). The entire equation of state has been found for the first time and has been shown to be very similar to the bulk. The melting temperatures have been shown to match much more closely and over a density range twice that originally reported. (See the caption to Fig. 5.) Although the order of the film phase transition and the volume and entropy changes differ, the form of the transition is remarkably similar to bulk melting, as may be seen in Fig. 4.

The above correspondences are strong evidence that the observed film behavior is fundamentally related to the helium-helium potential and dimensionality, and is not significantly obscured by the periodic potential of the Grafoil or by substrate inhomogeneity. (19) In the final section, we will discuss two recent theories about this melting behavior.

4. MELTING THEORIES

Two theories of two-dimensional melting have recently been offered, one by Dash and Bretz, (20) the other by Kosterlitz and Thouless. (21) In collaboration with the project presented here, Feynman (22) had independently adopted the point of view and confirmed the results of Kosterlitz and Thouless before we discovered their papers. He has urged us to call attention to their work. Although neither type of theory is yet in a form designed to be specifically applicable to the helium case, it is instructive to compare the theories to each other and to the data at hand.

These theories not only differ in the details of the melting phenomenon, but also arise from fundamentally contrasting views of the nature of crystallinity and ordering in two dimensions. It is well known that for a two-dimensional solid with short-range forces, correlations between the positions of atoms far from each other are washed out at nonzero temperature. (23) In the picture presented by Dash and Bretz, the melting of the two-dimensional solid is a consequence of the disappearance of these long-range correlations, which are always present in three-dimensional crystals. What they propose is a generalization to two dimensions of the empirical Lindemann formula, which may be deduced by noting that solids tend to melt when the root mean square displacement of atoms from their lattice sites is some arbitrary fraction of the interatomic spacing.

Kosterlitz and Thouless, on the other hand, feel that the loss of long range spatial correlations is unimportant, and that melting arises from the spontaneous growth of dislocations in the crystal. We shall show below, we believe for the first time, that the dislocation theory of melting in two dimensions actually gives rise to the empirical Lindemann formula with no arbitrary constants.

Dash and Bretz argue that two-dimensional solids begin melting as soon as the temperature rises above absolute zero. Atoms a small number of lattice spacings apart are still well correlated, and on that scale the substance may still be considered a solid. At longer ranges, however, the correlations are lost, so on that scale, the substance can no longer behave like a solid. The essential property distinguishing the solid from a fluid is its ability to support transverse sound modes. They thus hypothesize that the two-dimensional solid has transverse sound modes only of wavelength shorter than the range over which the atomic positions are well correlated. As the temperature is raised, the range of the correlations diminishes, and transverse modes are successively lost. They propose that the loss of each mode increases the entropy by $k_B/2$ (Boltzmann's constant). The process continues until all of these modes have melted. The contribution to the heat capacity that results from this change in entropy reflects the density of states for transverse modes; in particular, the observed peak in the heat capacity does not necessarily represent the endpoint of the melting process, but rather occurs at a temperature where modes are being lost at a wavelength that corresponds to a peak (a van Hove singularity) in the density of states for transverse modes. The peak corresponds to the melting of the last mode only in a pure Debye model, with the density of states constant up to a sharp cutoff. In this case the peak reflects the sharp cutoff, but it is just there that the Debye model is most unrealistic.

More realistically, using the computed modes of a tcp lattice (15), a peak in the transverse-mode density of states occurs at a wavevector, k_p, given by

$$k_p\sqrt{a} = 2\pi/\sqrt{3}$$

where a is the area of the unit cell. In this instance, the Dash-Bretz picture predicts a peak in the heat capacity at a temperature T_p given by

$$T_p = \frac{\gamma}{2.19 + \log(\xi\sqrt{3})} \frac{mk_B}{8\hbar^2} a\theta_D^2 \qquad [1]$$

whereas the last transverse mode vanishes at temperature T_m,

$$T_m = \frac{\gamma}{2.19 + \log \xi} \frac{mk_B}{8\hbar^2} a\theta_D^2 \qquad [2]$$

Here ξ is a number of order 1, γ is the ratio $<\delta^2>/a$ where $<\delta^2>$ is the relative mean square displacement of nearest neighbors at melting, and the 2.19 is actually $\log 2\sqrt{2\pi}$ plus Euler's constant. The formula for T_m is just the well-known Lindemann formula for the melting of solids, and in fact, the Dash-Bretz model is designed to reduce to the Lindemann picture in three dimensions, where a is replaced by the two-thirds power of the molecular volume. Comparing T_p and T_m, we see that T_p is lower by about 20%.

In computing the range of the correlations that led to these formulae, Dash and Bretz have assumed that $T \gg \theta_D$, whereas the helium melting peaks occur in the opposite limit. Moreover, as the authors point out, there are internal inconsistencies and other questions in the simplified picture they present. In trying to find a point of comparison between theory and experiment, we note that their assumption of equal melting entropy per mode requires that the two-dimensional melting entropy be one-half the three-dimensional value, while we find from our absolute entropy measurements that the ratio is closer to one-third. This assumption, however, is somewhat ad hoc, and also probably makes sense only for $T \gg \theta_D$, where all the modes are fully excited. It is therefore difficult to draw any firm conclusions about the theory. The expressions for T_p and T_m can only be taken as rough indications. However, we shall return to them below.

In contrast to the Dash-Bretz picture, Kosterlitz and Thouless predict a true phase transition at a definite temperature, T_M. At finite temperature below T_M, the medium supports shear modes of all wavelengths because the free, unpaired edge dislocations that would be required to relieve shear stresses are precluded by the large (formally infinite) free energies their creation would require.

There is a kind of long-range order found in the medium below T_M, which can be characterized by an appropriate order parameter. Above T_M, the existence of free edge dislocations becomes energetically favorable (they are created thermally) and the shear modes vanish.

Kosterlitz and Thouless propose a topological order parameter designed to detect the presence of unpaired dislocations by counting the net Burgers vector in a given area. Feynman prefers a different order parameter which stresses the fact, noted by others, (23) that even though the two-dimensional crystal loses its long-range spatial correlations, it retains its angular correlations. Consider a vector connecting two neighboring atoms in the lattice at T = 0, which we call $\vec{C}_{oo}$, and another, parallel to it at T = 0, but n by m lattice spaces away, $\vec{C}_{nm}$. The order parameter is then $\langle \vec{C}_{oo} \cdot \vec{C}_{nm} \rangle$, where $\langle ... \rangle$ means thermal average. It is easy to show (24) that this order parameter, in an harmonic, two-dimensional solid, remains nonzero at finite T even as $n, m \to \infty$. Thus, although the spatial correlations are lost, the angular correlations remain, and in this sense, the two-dimensional solid has long-range order.

There are a number of ways to argue that the growth of dislocations leads to a phase transition in two dimensions. One of the most instructive is to examine the medium below T_M, and consider the effect on the shear modulus of the thermal excitation of dislocation pairs. The energy of a pair of dislocations of opposite Burgers vectors a distance r apart is

$$E_2 = 2 L \log r + \text{constant} .$$

where L depends on the elastic constants of the two-dimensional medium. By contrast, the energy in the strain field of an unpaired dislocation is proportional to log A, where A is the area of the medium, formally taken to be infinite. At low T, there are no unpaired dislocations to be found, but bound pairs, or dipoles, are thermally excited in equilibrium.

When a shear stress is applied, the dislocation pairs respond, both by polarizing and by stretching apart. These two effects, each proportional to r, contribute to decreasing the shear modulus (defined in the limit of zero applied stress). The material will therefore restore shear stresses only so long as $\langle r^2 \rangle$ is finite,

$$\langle r^2 \rangle \sim \int r^2 e^{-E_2/k_B T} r \, dr$$

$$\sim \int r^{2(1 - L/k_B T)} r \, dr$$

We identify as T_M the temperature at which $\langle r^2 \rangle$ becomes infinite,

$$T_M = L/2k_B .$$

Kosterlitz and Thouless give a formula for L in terms of the shear modulus and Poisson's ratio for the two-dimensional solid, based on a standard result of three-dimensional elastic theory. There is some ambiguity in the meaning of the two-dimensional Poisson's ratio, however. Feynman has recomputed L from first principles in two dimensions, casting the result in terms of the longitudinal and transverse speeds of sound, c_l and c_t,

$$L = \frac{mb^2}{4\pi a} \frac{c_t^2 c_l^2}{c_t^2 + c_l^2} .$$

where b is the Burgers vector, which we shall set equal to $\sqrt{a}$. The two necessary elastic constants have been measured by Stewart, Siegel, and Goodstein (16), or could be deduced from our own data, but they arise here in precisely the same combination as they do in the Debye temperature, θ_D. Thus, only θ_D is needed to predict T_M at a given density,

$$T_M = \frac{1}{(2\pi)^2} \frac{mk_B}{8\hbar^2} a\theta_D^2 . \qquad [3]$$

This result has the same form as the Lindemann formula, Eq. 2. We shall return to discuss this below.

At first glance, the dislocation model appears to predict a first order phase transition. This is because the transition temperature is proportional to c_t, or to the shear modulus, which therefore must change discontinuously to zero upon melting. However, the derivation we have given shows that this conclusion is not inevitable. As dislocation pairs appear and grow apart, c_t and the shear modulus are reduced by them with increasing temperature. This reduces the pair energy, E_2, causing more to appear, further reducing c_t, and so on. At some point, the process runs away and the phase transition occurs. Kosterlitz and Thouless have tried to investigate this feedback effect by mean field theory. The result is to depress the transition temperature somewhat, and leave the thermodynamic functions nonsingular at T_M. However, as they point out, mean field theories are notoriously undependable in giving details of phase transitions. We note that the data of Thomy and Duval for the melting of krypton on exfoliated graphite (2) seem to show a transition that changes from first to higher order with increasing temperature. Thus, the order of the transition in two dimensions may not even be unique.

In Fig. 5, we compare the observed positions of the melting peaks for ⁴He on Grafoil to the simple prediction of T_M in Eq. 3. Although the predicted dependence $T_M \sim a\theta_D^2$ is well obeyed, the predicted values of T_M are close to 3/4 of the observed peak temperatures. As we have just argued, we should have expected the theory to overestimate the transition temperature. The difficulty, however, may lie in the use of classical elastic theory, ignoring the influence of zero-point energy and initial stress in solid two-dimensional helium. Moreover, we have used $b^2 = a$ in Eq. 3, whereas, for tcp, $b^2 = 1.15\, a$, raising the predicted T_M.

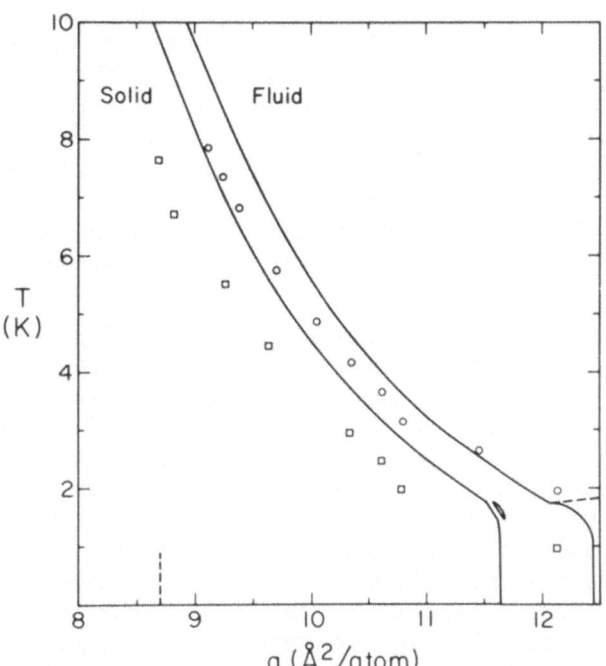

FIG. 5. The temperature of melting. The peaks in the experimental heat capacity (3) are shown by circles. The three circles above 6 K have been corrected to the coverage in the first layer alone, using the model of section 3. The squares show the predictions of Eq 3, using experimental θ_D (3). The solid lines show the phase diagram of bulk helium. In order to have the interatomic spacing identical in the hcp and tcp solids, we have set $a = 1.091\, v^{2/3}$. The same scaling was used in Figs. 2 and 4.

Eqs. 1 and 2 from Dash and Bretz have the same dependences, as well as adjustable parameters, and therefore may be made to agree quite well with the data. However, these formulae are derived for high temperature melting ($T_m \gg \theta_D$) and therefore are not the correct predictions of the Dash-Bretz theory for the helium case. Domb and Dugdale (25) have observed the same phenomenon in three-dimensional solid helium: the high temperature form of the Lindemann formula fits the melting curve quite well, but the presumably more correct form, based on the mean square displacements when not all modes are fully excited, does not agree with the data at all. This curious behavior may be accounted for if it turns out that the dislocation model produces the Lindemann formula in three dimensions as it does in two.

Let us return to comment upon the striking resemblance between Eq. 2 for T_m and Eq. 3 for T_M. It does not indicate that in some deep sense the two types of theory are the same; as we have already argued, they are quite different. There is, however, an important point to be made. As we have stated above, the empirical Lindemann melting formula can be derived by assuming that a crystal will melt when the mean square fluctuations in atomic positions reach some fraction of the interatomic spacing, i.e., when the parameter $\gamma = \langle\delta^2\rangle/a$ reaches some value, quoted by Dash and Bretz to be about 1/16 for ordinary substances. (20) (There is some dependence on crystal structure.) This fact underlies the Dash-Bretz theory, but the formula, Eq. 2, for T_m does not depend in any way on their picture of mode by mode melting; it is merely the point at which nearest neighbors in two dimensions satisfy the same basic melting criterion that all pairs, including nearest neighbors, satisfy in three dimensions. As such, it is entirely consistent with the Kosterlitz-Thouless-Feynman view that long-range spatial correlations are unimportant. More significantly, we have demonstrated here, we believe for the first time, that the Lindemann empirical formula can be derived from the dislocation picture of melting, at least in two dimensions, and in contrast to the atomic fluctuation picture, there is no empirical parameter. Comparing Eqs. 2 and 3, we see that we have predicted the Lindemann parameter,

$$\gamma = 2.19/(2\pi)^2 \cong 1/18 \ .$$

Although, as we can see from Fig. 5, this value is too low for helium, it is close to the value noted above for ordinary solids.

ACKNOWLEDGEMENTS

We wish gratefully to acknowledge the very substantial contributions made to this work by Professor R. P. Feynman. The final section of this paper had been written before Kosterlitz and

Thouless' most recent paper arrived. Little more than the names in the text had to be changed when we became aware of their prior work. We have also profited from many discussions with Jeffrey Greif and G. Alexander Stewart.

REFERENCES

1. Allan Widom, Phys. Rev. 185, 344-47 (1969). Statistical Mechanics of Ultrathin Films of Quantum Liquids.
2. André Thomy et Xavier Duval, J. Chim. Phys. Physiocochim. Biol. 67, 1101-10 (1970). N° 155.--Adsorption de Molécules Simples sur Graphite. III --Passage de la première couche par trois états successifs.
3. M. Bretz, G. B. Huff, and J. G. Dash, Phys. Rev. Letters 28, 729-31 (1972). Solid Phase of He⁴ Monolayers: Debye Temperatures and "Melting" Anomalies.
4. Trademark of Union Carbide Carbon Products Division, 270 Park Avenue, New York.
5. Robert L. Elgin, Ph.D. thesis, Caltech (unpublished, 1973). The Thermodynamics of the ⁴He Submonolayer Film Adsorbed on Grafoil.
6. L. D. Landau and E. M. Lifshitz, Statistical Physics, Pergamon Press Ltd. (London 1958).
7. Robert D. McCarty, U. S. Nat. Bur. Stand. Tech. Note 631 (1972). Thermophysical Properties of Helium-4 from 2 to 1500 K with Pressures to 1000 Atmospheres.
8. Michael Bretz and J. G. Dash, Phys. Rev. Letters 26, 963-65 (1971). Quasiclassical and Quantum Degenerate Helium Monolayers.
9. D. C. Hickernell, E. O. McLean, and O. E. Vilches, Phys. Rev. Letters 28, 789-92 (1972). Very Low-Temperature Specific Heat of Submonolayer Helium Films.
10. R. L. Elgin, G. A. Stewart, and D. L. Goodstein (to be published).
11. O. V. Lounasmaa, Cryogenics 1, 212-21 (1961). D. Q. Edwards and R. C. Pandorf, Phys. Rev. 140, A816-25 (1965). J. S. Dugdale and J. P. Frank, Phil. Trans. Roy. Soc. 257, 1-29 (1964).
12. Anthony D. Novaco, Phys. Rev. A 7 (May 1973 to be published). Cluster Expansion for Superlattice Physisorption: Helium Adsorbed upon Graphite.
13. Michael Bretz, private communication.
14. W. R. Gardner, J. K. Hoffer, and N. E. Phillips, Phys. Rev. A 7, 1029-43 (1973). Thermodynamic Properties of ⁴He. The hcp Phase at Low Densities.

15. P. Dean, Proc. Camb. Phil. Soc. $\underline{59}$, 383-96 (1963).   The vibrations of three two-dimensional lattices.

16. G. A. Stewart, S. Siegel, and D. L. Goodstein, Proc. 13th Int. Conf. Low Temp. Phys., Boulder, Colo. 1972.   Elastic Properties of Solid He^4 Monolayers from $4.2°K$ Vapor Pressure Studies.

17. R. L. Elgin and D. L. Goodstein, Proc. 13th Int. Conf. Low Temp. Phys., Boulder, Colo. 1972.   Thermodynamic Functions for Helium 4 Submonolayers.

18. M. Bretz, J. G. Dash, D. C. Hickernell, E. O. McLean, and O. E. Vilches, Phys. Rev. A $\underline{8}$ (1973 to be published).   Phases of He^3 and He^4 Monolayer Films Adsorbed on Basal Plane Graphite.

19. It is possible to rule out major inhomogeneity effects on the melting peak, such as the smearing out of an intrinsically first order phase transition. For more discussion of inhomogeneity and other parts of the phase diagram, see references 5 and 10.

20. J. G. Dash and M. Bretz, J. Low Temp. Phys. $\underline{9}$, 291-306 (1972). Short-Range Order and Melting Anomalies in Thin Films.

21. J. M. Kosterlitz and D. J. Thouless, J. Phys. C: Solid State $\underline{5}$, L124-26 (1972); $\underline{6}$, 1181-1203 (1973).   Long range order and metastability in two dimensional solids and superfluids; Ordering, metastability and phase transitions in two-dimensional systems.

22. R. P. Feynman, private communication.

23. N. D. Mermin, Phys. Rev. $\underline{176}$, 250-54 (1968).   Crystalline Order in Two Dimensions.

24. Some of the dependences of this order parameter have been computed for us by Jeffrey M. Greif (private communication). See also reference 23.

25. C. Domb and J. S. Dugdale, in: C. J. Gorter, ed. Progress in Low Temperature Physics, vol. II, 338-67, North-Holland Publishing Company (Amsterdam 1957).   Solid Helium.

 * Work supported by the National Science Foundation and a grant from the Research Corporation.

SOME ASPECTS OF QUANTUM THEORY OF ADSORBED HELIUM FILMS

Frederick J. Milford

Battelle Memorial Institute

Columbus, Ohio 43201

INTRODUCTION

The main purposes of this paper are to provide background
information for the theoretical papers that follow by Schick,
Novaco and Woo and to review some recent developments in areas
that these papers will not cover.

THE INTERACTION OF HELIUM ATOMS WITH A SOLID SURFACE

In considering a quantum theory of adsorbed helium films,
one of the first steps is to determine the interaction of helium
atoms with the adsorbing surface. There are several kinds of
surfaces that must be considered and the situation for each of
these is somewhat different. Three of the more important kinds
of surfaces are rare gas solids, metals, and graphite. These
are considered in the following.

Rare Gas Solids

Rare gas solids consist of relatively well known arrays of
rare gas atoms. The rare gas atoms are little affected by their
binding into a solid and in consequence it is a reasonable first
approximation to determine the interaction of a helium atom with
a rare gas solid by summing the two body potentials between the
atoms in the rare gas solids and the helium atom. In order to do
this it is necessary to know what the two body interaction is. It
has become customary to use the Lennard-Jones potential to describe
this two-body interaction. This procedure has some shortcomings

since it is well known that the Lennard-Jones potential is deficient
in several ways. It would be feasible to use a more accurate
numerically specified potential, especially since the final helium-
surface potential is specified numerically. At the present time
the lack of direct experimental knowledge of the helium-surface
potential makes the value of such a procedure questionable. Furthermore, the relatively consistent use of the Lennard-Jones potential facilitates the intercomparison of calculations. The main
problem is determining what parameters to use to characterize the
Lennard-Jones potential. Two procedures have been used. In one
of these the Lennard-Jones parameters have been obtained by combining those for the pure rare gas with those of helium by taking the
square root of the product of the well-depth parameters and half
the sum of the sigma (range) parameters. This procedure has some
theoretical basis but it should be considered mainly as an arbitrary
but reasonable procedure. The alternative is to take the Lennard-
Jones potentials obtained by fitting scattering data. These parameters tend to be appropriate to close approaches and so emphasize
the region of the potential roughly between its minimum and the
hard core radius. The differences here are another reflection on
the difficulty in determining an adequate interaction potential.
This poses even greater difficulties in the case of the helium-
helium interaction as discussed later in this paper.

Once the two body potential has been established it is a
simple matter, at least with the availability of a large scale
computer, to sum this potential over the relatively well determined
array that characterizes the rare gas solid. The nearest 100 atoms
is usually adequate. This produces a numerical potential, which
in some ways is unfortunate, but the difficulties that have arisen
in the past by approaches which seek analytical approximations are
sufficient to make it preferable to accept the complexity of a
numerically valued potential.

It is clear that in the case of a helium atom interacting
with a rare gas solid, there may be many body effects. That is,
the polarization of any given atom in the rare gas solid represents
a response to the total electric field at the site of that atom.
This electric field is produced not only by the oscillating helium
dipole but also by the oscillating dipoles of all of the other
atoms in the rare gas solid and particularly those components which
are induced by the helium atom. This problem can be treated, and
in fact has been treated, by at least two groups [1,2] of workers.
The net result is that for helium adsorbed on any rare gas solid
the total correction from this source as compared to the Lennard-
Jones potential is less than two and one-half percent. Consequently
it is felt that at least at this time this correction is not
important.

The Interaction of Helium Atoms with Metal Surfaces

When the interaction of helium atoms with metal surfaces is considered, the situation is altogether different. There are a number of older calculations ranging from those as simple as a consideration of the induced image dipole and its interaction with the helium atom to more complicated calculations that attempt to take into account the actual nature of the electron distribution in the metal. The difficulties with these calculations are well known and much discussed. None of these calculations uses a realistic model for the surface; that is, they all use a semi-infinite solid with the same properties as the bulk metal. Secondly, the repulsive potential as the helium atom approaches very close to the metal surface has been considered only by Pollard and his calculation has a variety of shortcomings. In all of the older calculations the origin for z is a plane passing through the first layer of ion cores and this plane is typically the image plane for large separations between the helium atom and the metal surface. It is clear that a more appropriate image plane would be half the interplanar spacing outside the plane of the ion cores. Also recent calculations directed towards the consideration of chemisorption of hydrogen [11,12,13] indicate that the image plane is displaced even further away from the plane of the ion cores. Admittedly in this case there is an extra electron added to the metal and this might be expected to displace the image plane, but it is nonetheless clear that this represents a definite shortcoming in the older calculations.

Our knowledge of the interactions of helium atoms with metal surfaces is poor from both theoretical and experimental viewpoints. From the theoretical standpoint a possibly promising approach is the inhomogeneous electron gas formalism mentioned above. In view of the striking success obtained with chemisorption, this formalism may make possible a realistic calculation of the repulsive interaction and it may further yield at least the form of the long range part of the attractive interaction. The latter, of course, is more of a problem since it depends on the dynamics of the conduction electrons which are not at present included in electron gas formalism. There is a considerable body of work on scattering and on thermal desorption but these experimental studies have not produced satisfactory characterization of the interaction of helium with metal surfaces. In fact, some of the scattering experiments, in particular the scattering of helium from single crystal metal surfaces, are rather complete mysteries.

The Interaction of Helium Atoms with Graphite

The adsorption of helium on graphite has been studied for many years. Earliest studies involved adsorption on high surface

areas of graphitized organic materials. More recently it has been
recognized that exfoliated graphite provides almost plane exposure
of large areas of basal planes of graphite and consequently that
this material represents a very interesting substrate for the
study of adsorbed helium.

Substantially all attempts to construct a potential describing the interaction of helium with graphite have used the Lennard-
Jones form of the potential and a simple summation over the semi-
infinite array of graphite carbon atoms. The Lennard-Jones parameters have been obtained by interpolation as indicated above for
rare gas solids. There seems to be no scattering data available
from which an alternative set of parameters might be obtained.
The potential so obtained is somewhat different from that characteristic of rare gas solids in that there is a strong attractive
well in the z direction but only a rather weak periodic potential
in the x and y directions.

The problems with this potential are two. On the one hand,
the Lennard-Jones parameters characterizing the carbon helium
interaction are not known directly. On the other hand covalency
in the graphite is ignored. In spite of these shortcomings we
feel that a reasonable potential has been found. Comparison with
existing experimental work on heats of adsorption, for example,
produces agreement to within about 20%. It should be noted that
the periodic potential is almost certainly overestimated in this
calculation and that the calculated periodic potential is only
about 10% of the total well depth. It thus seems reasonable to
think of a plane parallel to a graphite basal plane as an almost
smooth adsorption surface and consequently to approach further
calculations at least in first approximation as particles that are
bound in the z direction and essentially free in the other two
directions.

SINGLE PARTICLE STATES

Having determined interaction potential, single particle
states are determined by solving the Schrödinger equation with
that potential. This program has several problems. On the one
hand the potential is not separable, that is it does not separate
into the sum of the part which depends on x and y, another part
which depends on z. This results in computational complexities
which one might very well describe as quantum chemical complications since the same kinds of problems plague that field. On the
other hand the potential is periodic in two variables. In consequence of this the Floquet or Bloch theorem applies and the energy
eigenstates are bands. The eigenvalues are labeled by a continuous

variable k, a band index i, and an integer n, which characterizes
excitation perpendicular to the surface. For practical purposes
k and i suffice. The z excitations are extremely high in energy
and consequently they play a very small role in low temperature
experiments.

Calculations have been done in considerable detail for rare
gas solids, for rare gas plated copper, graphite, and rare gas
plated graphite.* In some cases these calculations have been done
by several groups and the results in general compare extremely well.
From the computational standpoint and talking only about single
particle states, we seem to be confronted with a problem which is
relatively completely solved and relatively well understood.

The results of these calculations may be compared with experiments in two general ways. First the isosteric heat of adsorption
at zero coverage is just the binding energy of the helium atom in
the lowest state. The isosteric heat of adsorption has been measured for helium on argon plated copper by Daunt and Lerner [10]
who obtained a value between 125 and 150°K. Another measurement
by Pollack et al. [14] using a diffusion technique yields 102°K
for the binding energy. These compare with a value of 60° calculated along the lines outlined above [3]. The comparison seems to be
rather poor and perhaps the experimental work or the calculations
would bear some repetition. The case of helium on graphite,
Grayson and Aston, in some old work, obtained the value of 151°K
for the isosteric heat extrapolated to zero coverage. Corresponding calculations yield a value of 189°K [7], somewhat better agreement than in the case of helium on argon plated copper. Very
recently, Lee and Gowland [9] have looked at the desorption of
helium from an argon surface. They obtain an adsorption energy of
150°K. While this adsorption energy may not be exactly the energy
of the lowest bound state for helium adsorbed on an argon surface,
it is very closely related and they are probably the same within
about 5%. The corresponding calculations give 170°K. The binding
of helium atoms on copper has been measured by an ingenious use of
modern technology to implement the old idea of relating surface
binding to diffusion down a small bore tube [14]. The result,
177°K, cannot, however, be compared with theoretical calculations
since none of the latter have been made. It is interesting to note
that except in the case of helium on argon plated copper, experimental results are always smaller than the calculated values, but
the significance of this observation is not clear.

The original motivation for many of the studies of interactions
of helium atoms with solid surfaces and computations of single
particle states was to contribute to the understanding of the heat

* Running time is of the order of 5-10 min. on a CDC 6400.

capacity data being generated primarily by Dash and co-workers. To
this end the heat capacity has been calculated for non-interacting
helium atoms at low coverages. Unfortunately, there seems to be
little connection between these calculations and the experiments.
It seems highly probable that interactions between helium atoms
must be considered in order to explain the experimental data.

INTERACTION POTENTIALS AND PAIRING

The two helium atoms adsorbed on a solid surface interact with
one another. This interaction may lead to the formation of the
bound pair, and if many atoms are adsorbed this interaction potential will lead to condensation and to a two dimensional liquid or
even to the formation of a two dimensional solid. The question of
the existence of such phases is one of the more prominant features
of contemporary research and a careful investigation of the interaction seems appropriate at this stage.

Interaction of Two Helium Atoms

The interaction of two helium atoms has been studied both
experimentally and theoretically for almost half a century. In
spite of all our refined contemporary knowledge of the situation,
it is worth noting that the Lennard-Jones potential which appeared
in print in 1924 is still very commonly used. It is also worth
noting that the deBoer-Michaels parameters, which were determined
by fitting virial coefficients were proposed in 1938 and are still
commonly used in the Lennard-Jones potential for helium-helium
interaction. Finally, in spite of this long history of studies,
the potential is still uncertain in some ways. There are several
approaches to exploring the helium-helium interaction approaches.
They may be characterized as experimental in which one starts with
experimental data and attempts to invert it to determine the potential that is responsible for this data. A great deal is known
about the feasibility of this process for scattering data and in
particular, it is clear that what is needed is a double differential
scattering cross section for all energies and all angles. Such data
is not yet available and consequently the interaction potential has
not been determined in this relatively direct experimental way.
The situation is, however, improving but that was the situation with
the proton-proton scattering cross section many years ago. People
kept saying if only we knew the scattering cross section for higher
energies we could determine the potential exactly. Eventually they
did and the high energy explosion was upon us. Perhaps the same
thing will happen in the helium-helium scattering experiments, but
it seems unlikely.

An alternative to this approach is to characterize a reasonable form for the potential by a small number of parameters, and then to fit the experimental data by adjusting the parameters. The classic case is, of course, the use of the Lennard-Jones potential, but many others have also been used. The difficulties with this approach are primarily that the number of parameters is usually sufficiently small that the potential that is obtained fits one set of experiments but not another, at least not without using a different set of parameters. Some experiments are sensitive to the characteristics of the potential at short range and some to the region of the minimum and others to the tail. It is important to use parameters that fit the potential well in the regions that are important in the problem being considered. The third approach to determine a potential is the so-called ab initio approach which basically means starting from the Schrödinger equation for two protons and four electrons. The Born-Oppenheimer approximation and the Heller-Feynman theorem are usually invoked to separate the nuclear motion from the electronic motion on the one hand and to obtain the interaction potential from electrostatic consideration after the electron distribution has been determined. A great deal of progress has been made in this area in recent years.

These questions are sufficiently important that it seems appropriate to mention briefly some specific calculations. First of all from the ab initio standpoint Bertoncini and Wahl [19] on the one hand and Schaefer et al. [20] on the other, have presented extensive calculations on the short range and intermediate range interaction of helium atoms. The calculations agree very precisely and represent a definite advance in the state of the art. The overall situation regarding the exact quantum mechanics of atomic interactions has been reviewed recently by Schaefer in a significant book. Another paper worthly of mention is the calculation by Gordon and Kim [15] of the interaction of rare gas atoms. Unfortunately their work on helium seems to be at gross variance with the experimental situation.

Recent experimental work using cross beam scattering by Siska et al. [18] has fitted an eight parameter potential. This appears to give a reasonable approximation to the interaction potential over the entire range of distances, but eight parameters is probably too many to use for most purposes. Another recent scattering experiment has been the very low energy work done by Oates and King [17]. They obtained the total cross section as a function of velocity of incident particle. Their measurements were compared with empirical potentials including the Lennard-Jones and Slater potentials. The experimental results lie between those predicted by these two potentials. It is interesting also to note that their cross section shows some structure and that this structure may be significant. None of the theoretical predictions show structure of any kind.

On the basis of these considerations, it seems appropriate to
note that the value of the distance at which the potential is zero,
commonly known as sigma, varies by about 5% among the various proposed potentials. The depth of the potential well varies about 1°K
which is about 10% of the total depth. This leads to the obvious
comment that one should be rather careful in calculating small
energies using potentials as uncertain as this.

The Influence of the Substrate on the Interaction

There are basically two ways in which the adsorbing surface
can affect the interaction between two helium atoms, namely through
the electronic distribution and through the phonon distribution.
Considering the electronic phonomenon first it is very simple to
see that in the case of a metal, very crudely, each dipole, that is
each of the oscillating dipoles that represent a polarized helium
atom, produces an image in the metal. A very elementary consideration of the interaction of the two helium atoms with the two
corresponding images shows that there may be a reduction of as
much as one-third in the interaction. In the case of insulators,
the same phonomenon occurs. The calculation is more complicated
but it is clear that here too, there may be substantial effects.
There have been several calculations of these effects starting with
Sinanoglu and Pitzer. All of these calculations use a very poor
approximation in the surface structure. That is, they use a semi-
infinite solid with the properties of bulk materials. In the case
of insulators this is perhaps better justified than in the case of
metals. Recently the effect of many body interactions on the interaction of two helium atoms adsorbed on a rare gas solid has
been considered. These calculations have, however, been restricted
to the interaction on the heavier rare gas atoms.

Phonon effects have been considered by Schick and Campbell [16].
The basic process is the exchange of the surface phonon between two
adsorbed helium atoms. This produces an attractive interaction
much as in the case of the electron-phonon situation. The result
of the calculations done by Schick and Campbell indicate an augmen-
tation of the interaction by about 1°K, that is about a 10% effect.
In view of other uncertainties, it seems that for the time being,
this effect can be ignored.

The Binding of Pairs of Adsorbed Helium Atoms

The binding of pairs of helium atoms has been considered in
two ways. On the one hand, Bagchi [23] has considered a strictly
two dimensional problem in which helium atoms in two dimensional
plane wave states interact through Lennard-Jones potential. This

is a straight forward but complicated calculation that leads to
very small binding energies, a few hundredth of a degree Kelvin in
the case of Helium four and slightly more than $10^{-7}°K$ in the case
of Helium three. It seems reasonable to speculate that if motion
in the third direction were taken into account the binding would
be even weaker, since it would be expected to be somewhere between
Bagchi's values and the unbound situation for two helium atoms
free to move in three dimensions. Bagchi also considered a tight
binding approximation in which he considered the pairing of helium
atoms on a two dimensional substrate characterized by square potential wells and tunneling times. The calculation is enormously
more complicated than the free particle calculation. Bagchi finds
that under some circumstances there may be a bound state for two
helium atoms adsorbed on such a substrate, however, it is difficult
to relate this calculation to the precise situation of helium atoms
adsorbed on a solid surface. The difficulty stems from two sources,
first Bagchi's choice of a rectangular potential well and secondly,
the fact that both of his calculations are strictly two dimensional.
His conclusion, namely that the presence of a surface appears to
tend to promote the binding of two helium atoms adsorbed on a solid
surface, does, however, seem reasonable.

The other calculation of the binding of adsorbed atoms makes use
of single particle wave functions appropriate to the specified adsorbing surface introduces a correlation and it calculates the expectation value of the interaction energy. It is found by Lai, Woo and
Wu [22] that a pair of helium atoms is bound by about 3°K. Cole [24]
has objected to this calculation because of the omission of the factor of 1/N in the normalization which leads to zero binding energy
per pair. However, as Cole points out with a pair wave function
which vanishes for infinite separation of the pair there may be binding and in that case the result obtained by Lai, Woo and Wu may, in
fact, be correct.

In any case, it seems that the most that can be said reliably
at the present time is that the presence of a surface may enhance
the binding of a pair of helium atoms.

ACKNOWLEDGEMENT

The author is indebted to Dr. S.-T. Wu for helpful discussions
and other assistance in the preparation of this paper.

REFERENCES

Substrate Potential and Single Particle States (c.f. also Ref. 22)

1) W. M. Gersbacher, Jr. and F. J. Milford, J. Low Temp. Phys. 9,
    189 (1972).

2) T. B. MacRury and B. Linder, J. Chem. Phys. 54, 2056 (1971).

3) F. J. Milford and A. D. Novaco, Phys. Rev. A 4, 1136 (1971).

4) A. D. Novaco and F. J. Milford, Phys. Rev. A 5, 783 (1972).

5) D. E. Hagen, J. Chem. Phys. 56, 5413 (1972).

6) R. Dovesi, E. Garrone, and F. Ricca, Atti Accad. Sci. Tovino I (Italy) 105, 831 (1971).

7) D. E. Hagen, A. D. Novaco, and F. J. Milford, Adsorption-Desorption Phenomena (F. Ricca, Ed.), Academic Press, New York, p 99 (1972).

8) F. Ricca, C. Pisani, and E. Garrone, Adsorption-Desorption Phenomena (F. Ricca, Ed.), Academic Press, New York, p 111 (1972).

9) T. J. Lee and L. Gowland, Adsorption-Desorption Phenomena (F. Ricca, Ed.), Academic Press, New York, p 137 (1972).

10) J. G. Daunt and E. Lerner, Adsorption-Desorption Phenomena (F. Ricca, Ed.), Academic Press, New York, p 127 (1972).

11) S. C. Ying, J. R. Smith, and W. Kohn, J. Vac. Sci. & Tech. 9, 575 (1972).

12) J. R. Smith, S. C. Ying, and W. Kohn, Phys. Rev. Letts. 30, 610 (1973).

13) N. D. Lang, Advances in Solid State Physics (Ehrenreich, Turnbull, & Sietz, Eds.), Academic Press, New York (to be published).

14) F. Pollack, H. Logan, J. Hobgood, and J. G. Daunt, Phys. Rev. Letts. 28, 346 (1972).

Interaction Potential

15) R. G. Gorden and Y. S. Kim, J. Chem. Phys. 56, 3122 (1972).

16) M. Schick and C. E. Campbell, Phys. Rev. A 2, 1591 (1970).

17) D. E. Oates and J. G. King, Phys. Rev. Letts. 26, 735 (1971).

18) P. E. Siska, J. M. Parson, T. P. Shafer, and Y. T. Lee, J. Chem. Phys. 55, 5762 (1971).

19) P. Bertoncini and A. C. Wahl, Phys. Rev. Letts. 25, 991 (1970).

20)   H. F. Schaefer, III, D. R. McLaughlin, F. E. Harris, and B. J. Alder, Phys. Rev. Letts. 25, 988 (1970).

21)   M. Cavallini, L. Meneghetti, G. Scoles, and M. Yealland, Phys. Rev. Letts. 24, 1469 (1970).

Pairing

22)   H.-W. Lai, C.-W. Woo, and F. J. Wu, J. Low Temp. Phys. 5, 499 (1971).

23)   A. Bagchi, Phys. Rev. A 3, 1133 (1971).

24)   M. W. Cole, Phys. Lett. A 38, 281 (1972).

25)   W. A. Steele and E. J. Derderian, Adsorption-Desorption Phenomena (F. Ricca, Ed.), Academic Press, New York, p 85 (1972).

26)   R. H. Paramenter, Phys. Rev. B 1, 1070 (1970).

SECOND VIRIAL COEFFICIENTS OF TWO-DIMENSIONAL SYSTEMS OF He^3 AND He^{4*}

R. L. Siddon and M. Schick

Department of Physics, University of Washington

Seattle, Washington 98195

ABSTRACT

Second virial coefficients of two dimensional systems of He^3 and He^4 are calculated from 0.1K to 60.0K assuming a Lennard-Jones 6-12 interaction. The resultant specific heat is compared with measurements of helium monolayers adsorbed on graphite. The results strongly suggest that the usual peaks in the low-density heat capacity can be attributed solely to two particle interactions.

Recent experiments have shown that low-coverage helium mono-layers absorbed on graphite behave as two-dimensional gases [1,2]. This behavior is indicated by the high temperature ($\sim$ 4K) approach of the specific heat to a value which is approximately Boltzmann's constant k, the value appropriate to an ideal gas. The deviations from ideal behavior are significant however. In particular, where-as the specific heat of He^3 monolayers decreases with decreasing temperature, that of He^4 monolayers shows a pronounced peak at temperatures near 1K. This difference cannot be attributed to the band structure induced by the periodic substrate potential [3]. Instead, it has been variously interpreted as indicating the presence of long-range inhomogeneities in the substrate [4,5] or of a liquid-gas transition which occurs in the boson system only. [6] In view of the recent spate of experiments which demonstrate the existence of superfluidity in He^4 films of only a few layers [7] it might be thought that the specific heat observed in the sub-monolayer He^4 reflects a transition to a superfluid state. That the existence of such a state is not necessarily excluded by the well-known theorem of Hohenberg [8] has been recently re-emphasized by Kosterlitz and Thouless [9].

However, the possibility remains that the data do not primarily reflect properties of the substrate or many-body effects
within the film but can be attributed to deviations from ideal behavior arising from two-particle interactions. Such deviations
are most directly expressed by the virial expansions of the thermodynamic functions. The second virial coefficients of two-
dimensional systems of He^3 and He^4 have been calculated by approximation methods (10), but these methods are unreliable at temperatures of order 10K and the results are therefore not applicable to
this problem. We have therefore calculated directly the required
second virial coefficient and the concomitant thermodynamic functions.

The second virial coefficient B of a two-dimensional system
can be expressed in terms of the phase shifts $\delta_m(E)$ and bound-state
energies E_m^B of the two-particle interaction $V(r)$ by means of a
procedure analogous to that employed in the three-dimensional case.
(11) For spinless bosons (fermions) of mass M, the expression is

$$B(\beta) = \mp \lambda^2/4 - 2\lambda^2 \Sigma(e^{-\beta E_m^B}-1) - (2\lambda^2\beta/\pi) \int_0^\infty \Sigma\delta_m(E)e^{-\beta E}dE \qquad (1)$$

where $\beta = 1/kT$, $\lambda^2 = 2\pi\hbar^2 \beta/M$, the upper (lower) sign is taken in
the first term, and the sums are over all even (odd) values of the
azimuthal quantum number m. The virial coefficient for spin 1/2
fermions is equal to the sum of 1/4 of the spinless boson coefficient and 3/4 of the spinless fermion coefficient. Once B is known
as a function of β, we can calculate the linear terms in a low-
density expansion of the entropy and specific heat at constant area.
With the areal particle density denoted by n, these expansions are
given by

$$S/Nk-S_0/Nk = -n(B-\beta dB/d\beta) + O(n^2), \qquad (2)$$

where $S_0/Nk = 2-\log [n\lambda^2/(2s+1)]$, the entropy per particle in
units of k of a classical ideal gas of spin s, and

$$C/Nk = 1-n\beta^2 d^2B/d\beta^2 + O(n^2). \qquad (3)$$

The interparticle interaction was taken to be the Lennard-
Jones 6-12 potential $V(r) = 4\varepsilon[(\sigma/r)^{12} - (\sigma/r)^6]$ with the de Boer-
Michels parameters[12] for helium of $\varepsilon/k = 10.22K$, $\sigma = 2.556\text{Å}$. The
phase shifts $\delta_m(E)$ and phase shift sums $\Sigma\delta_m(E)$ were calculated for
values of the dimensionless wavevector $q = (M\sigma^2E/\hbar^2)^{\frac{1}{2}}$ from 0.1 to
20.0 in intervals of 0.1. The behavior of the m equal to zero
phase shift with decreasing q indicated the presence of a bound
state for both He^3 and He^4, in agreement with Bagchi[13]. We
adopted his values of E_0^B which are -0.0236K and $-4.6 \times 10^{-7}K$ for
He^4 and He^3 respectively. The virial coefficients were calculated
for temperatures between 0.1K and 60.0K by carrying out the

integration in Eq(1) numerically. The error introduced by trun-
cating the integration at an energy corresponding to a q of 20.0
was investigated and found to be completely ignorable. Our results are tabulated in Table 1 and displayed in Fig. 1. Also
shown is the virial coefficient obtained classically. The difference between the He^3 and He^4 coefficients above 7K is due almost
completely to the mass difference. Below this temperature, the
effects of spin and statistics are also important.

The difference between the entropy per particle and that of
an ideal classical gas was calculated according to Eq. (2). In
Fig. 2, the negative of this quantity is plotted versus areal density for a He^4 system at temperatures of 13.62K, 7.87K, and 5.00K.
The points shown were supplied by R. L. Elgin who calculated these
values from the data of experiments on He^4 adsorbed on graphite
conducted by the California Institute of Technology and University
of Washington groups. (14) The increase in the values at densities $n \gtrsim .03(Å)^{-2}$ reflects the appearance of anamalous behavior
of the helium at very low coverages[2]. In the density range from
.03 to $.09(Å)^{-2}$ (monolayer completion density is $.115(Å)^{-2}$), the
data fall on a straight line. The major part of the deviation
of the entropy from ideal gas values is accounted for by our result.

Figure 3 shows the comparison between the values of the specific heat calculated according to Eq. (3) and experimental data (15)
for He^3 at densities of .0279 and $.0415(Å)^{-2}$ and He^4 at .0273 and
$.0399(Å)^{-2}$. It is seen that the agreement for the He^3 system is
rather good for all temperatures shown. The density dependence of
the specific heat is also given correctly by the linear term in
the virial expansion. Even more striking is the comparison between the calculated and observed values of the He^4 specific heat.
It is seen that the specific heat is predicted to rise above unity
below 2.26K and that, for different densities, the specific heats
should cross one another at this temperature. Both the rise and
crossing are observed experimentally, the latter feature occurring
at the predicted temperature. Neither the magnitude nor the position of the observed peaks can be obtained from the second virial
coefficient alone. The eventual decrease of the specific heat to
zero is contained in higher order terms of the virial expansion.

In order to determine the region of applicability of our results, which ignore these higher order terms, we re-examine the
experimental data in the form of a plot of $(C/Nk-1)n^{-1}$ versus
temperature. If, in Eq. (3), terms of order n^2 can be ignored,
then all data plotted as above should fall on a universal curve.
The He^3 and He^4 data are presented in this fashion in Figs. 4 and
5 along with the theoretical results. (16) It can be seen from
Fig. 4 that, in the density range investigated, the contribution

Table I. Second virial coefficients for two-dimensional systems of He^3 and He^4

HE^3

T	B	$B\frac{dB}{dT}$	$B^2\frac{d^2B}{dT^2}$
.10000	24.08147	76.03395	14.28803
.20000	-9.73568	28.11249	23.75279
.40000	-18.05412	1.62483	24.88358
.60000	-17.71666	-6.72439	20.99854
.80000	-15.29034	-9.79718	17.13717
1.00000	-12.95903	-10.93862	13.99543
1.20000	-10.43006	-11.23476	11.53833
1.40000	-9.70225	-11.13590	9.62662
1.60000	-7.73254	-10.85005	8.13182
1.80000	-6.67545	-10.48080	6.95307
2.00000	-5.39192	-10.07864	6.01444
2.20000	-4.45036	-9.67363	5.25930
2.40000	-3.62566	-9.27932	4.64539
2.60000	-2.90785	-8.90285	4.20105
2.80000	-2.25129	-8.52733	3.72240
3.00000	-1.67309	-8.21362	3.37128
3.20000	-1.15307	-7.90143	3.07342
3.40000	-.68290	-7.60078	2.81936
3.60000	-.25576	-7.33740	2.59960
3.80000	.13408	-7.08289	2.40808
4.00000	.49126	-6.84685	2.23973
4.20000	.81977	-6.62194	2.09055
4.40000	1.12295	-6.41288	1.95733
4.60000	1.40363	-6.21652	1.83752
4.80000	1.66427	-6.03177	1.72905
5.00000	1.90693	-5.85765	1.63026
6.00000	2.90657	-5.11861	1.24119
7.00000	3.65071	-4.54390	.96378
8.00000	4.22630	-4.08260	.75198
9.00000	4.68455	-3.70282	.58308
10.00000	5.05770	-3.38370	.44441
11.00000	5.36709	-3.11106	.32811
12.00000	5.62743	-2.87491	.22900
14.00000	6.04009	-2.48496	.06874
16.00000	6.35093	-2.17487	-.05540
18.00000	6.59198	-1.92136	-.15449
20.00000	6.80782	-1.72059	-.23543
22.00000	6.96370	-1.52963	-.30278
24.00000	7.06370	-1.37449	-.35949
26.00000	7.16877	-1.23917	-.40838
28.00000	7.25564	-1.11995	-.45049
30.00000	7.32923	-1.01400	-.48724
32.00000	7.39159	-.91914	-.51959
34.00000	7.44470	-.83365	-.54825
36.00000	7.49012	-.75615	-.57381
38.00000	7.52909	-.68554	-.59673
40.00000	7.56258	-.62091	-.61738
45.00000	7.62735	-.48080	-.66098
50.00000	7.67132	-.36482	-.69584
55.00000	7.70189	-.26705	-.72653
60.00000	7.72164	-.18337	-.74926

HE^4

T	B	$B\frac{dB}{dT}$	$B^2\frac{d^2B}{dT^2}$
.10000	-1134.80984	-1761.55188	-1195.24276
.20000	-401.72009	-584.42558	-307.12340
.40000	-150.06639	-207.85780	-97.01788
.60000	-86.43571	-115.38814	-49.76772
.80000	-58.17163	-76.98487	-28.08353
1.00000	-44.34907	-57.03024	-17.61925
1.20000	-35.08115	-45.20074	-10.82229
1.40000	-28.72871	-37.52406	-6.56985
1.60000	-24.08582	-32.19523	-3.84003
1.80000	-20.52989	-28.29711	-2.06059
2.00000	-17.76096	-25.32387	-.89107
2.20000	-15.40948	-22.97815	-.12077
2.40000	-13.49440	-21.07578	.38437
2.60000	-11.87161	-19.49764	.71146
2.80000	-10.47681	-18.16373	.91801
3.00000	-9.26365	-17.01854	1.04250
3.20000	-8.19782	-16.02242	1.11096
3.40000	-7.25330	-15.14631	1.14116
3.60000	-6.41000	-14.36846	1.14529
3.80000	-5.65213	-13.67222	1.13182
4.00000	-4.96706	-13.04460	1.10660
4.20000	-4.34461	-12.47537	1.07371
4.40000	-3.77641	-11.95627	1.03601
4.60000	-3.25558	-11.48060	.99549
4.80000	-2.77634	-11.04285	.95352
5.00000	-2.33386	-10.63843	.91107
6.00000	-.54758	-8.99996	.70822
7.00000	.74546	-7.80267	.53487
8.00000	1.72494	-6.88492	.39079
9.00000	2.49226	-6.15647	.27046
10.00000	3.10917	-5.56265	.16879
11.00000	3.61548	-5.06824	.08183
12.00000	4.03804	-4.64950	.00662
14.00000	4.70185	-3.97703	-.11696
16.00000	5.19769	-3.45897	-.21428
18.00000	5.58042	-3.04632	-.29291
20.00000	5.88336	-2.70906	-.35775
22.00000	6.12798	-2.42771	-.41212
24.00000	6.32871	-2.18908	-.45835
26.00000	6.49563	-1.98387	-.49811
28.00000	6.63597	-1.80532	-.53265
30.00000	6.75506	-1.64843	-.56292
32.00000	6.85693	-1.50937	-.58964
34.00000	6.94464	-1.38518	-.61340
36.00000	7.02060	-1.27355	-.63466
38.00000	7.08670	-1.17261	-.65383
40.00000	7.14448	-1.08084	-.67125
45.00000	7.26001	-.88399	-.70932
50.00000	7.34454	-.72293	-.74397
55.00000	7.40694	-.58791	-.78175
60.00000	7.45300	-.47184	-.83048

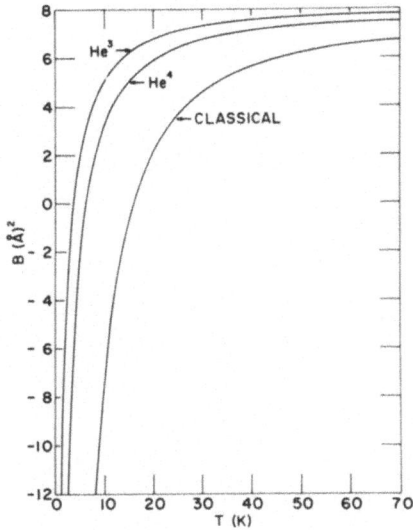

Fig. 1   Second virial coefficients of two-dimensional system of He³ and He⁴.   The classical result is also shown.

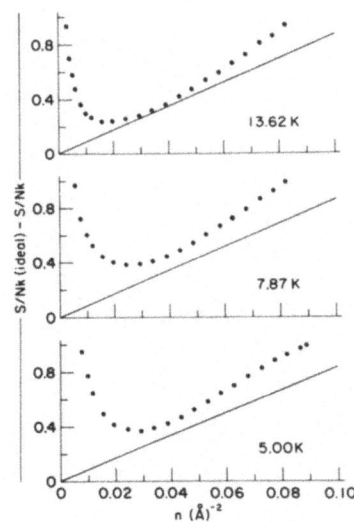

Fig. 2   Comparison between calculated and observed deviations of He⁴ entropy grom ideal gas values as a function of density.

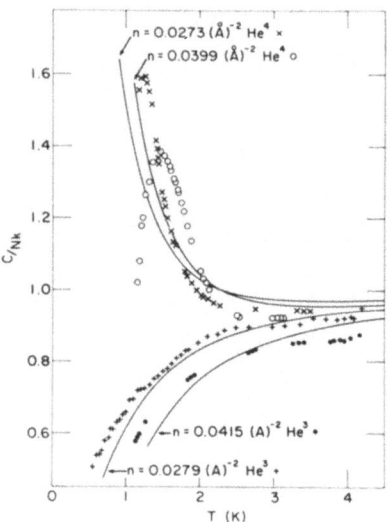

Fig. 3 Comparison between calculated and observed
values of the specific heat of He^3 and He^4 systems.

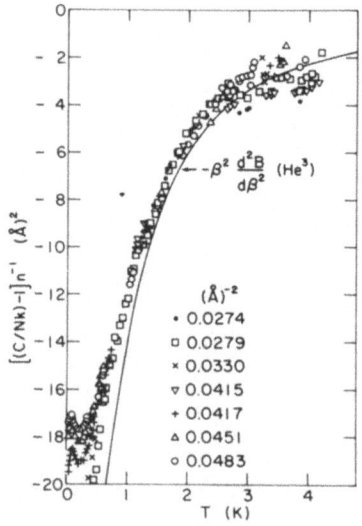

Fig. 4 Comparison between calcualted and observed
value of the specific heat of He^3. The plot makes mani-
fest the temperatures for which contribution to the
specific heat from the third virial coefficient and
higher can be ignored.

of the higher virial coefficients to the He^3 heat capacity can be ignored above temperatures of approximately 0.6K. Similarly, Fig. 5 indicates that the contribution of higher virial coefficients to the He^4 heat capacity can be ignored above 2.0K. This temperature range includes the region in which the heat capacity begins to increase with decreasing temperature. As C/Nk crosses unity at 2K, the contribution of higher order terms becomes apparent in Fig. 5. This is to be expected because, from Eq. (3), the contribution of the second virial coefficient vanishes as C/Nk approaches unity. Therefore the higher terms will be apparent eventhough they may be small.

We conclude from the above analysis that the appearance of the peaks in the He^4 heat capacity can be attributed solely to the two-particle interactions and do not necessarily signal long-range substrate inhomogeneities or a cooperative liquid-gas transition. It is to be emphasized, however, that the above results do not preclude the possibility of liquefaction.

It is interesting to consider the reasons for the great difference in the behavior of the He^4 and He^3 heat capacities. The second virial coefficient of the He^4 systems depends only upon those components of the interaction which, in a partial wave decomposition, are characterized by a z component of angular momentum equal to an even integer. In the temperature range of interest to us it can be shown that the δ_0 phase shift makes the largest contribution to the boson heat capacity. As collisions with m equal to zero sample predominantly the hard core repulsion, the rise in the He^4 heat capacity can be attributed to this part of the interaction.

The second virial coefficient of the He^3 system depends upon all components of the interaction, with odd m contributions weighted by 3/4 and even m contributions weighted by 1/4. When two particles scatter with a z component of angular momentum equal ±1, the average impact parameter of the scattering is of the order of the thermal wavelength λ. In the temperature range of interest to us, the thermal wavelength is of the same order of magnitude as the range of the attractive part of the potential. Therefore, scattering with m = ±1 is strongly affected by the attraction. These two contributions, weighted by 3/4, can be expected to dominate the single m = 0 contribution weighted by 1/4. To confirm this reasoning, we have calculated the contribution to the He^3 specific heat from each of the first few partial waves. The results are shown in Fig. 6. The curve labelled δ_1 is the sum of the contribution from δ_1 and δ_{-1} which are identical, and similarily for the other phase shift contributions.

The above analysis indicates that the difference in the

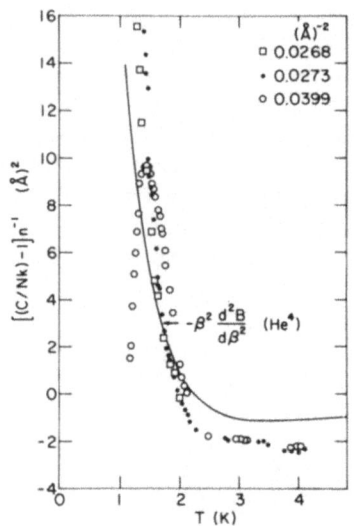

Fig. 5 Comparison between calculated and observed
values fo the specific heat of He[4]. The plot makes mani-
fest the temperature for which contributions to the
specific heat from the third virial coefficient and
higher can be ignored.

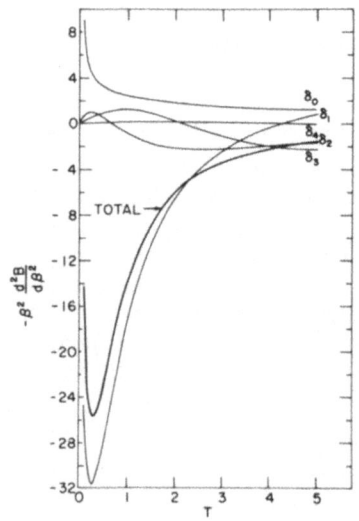

Fig. 6 Decomposition of the contribution of the second
virial coefficient to the He[3] specific heat according
to partial waves.

behavior of the specific heats is due to the fact that the repulsive part of the potential dominates the He^4 specific heat while the attractive part dominates the He^3 signal. However, this dominance of the He^3 signal by the attractive part of the interaction cannot persist to very low temperatures. This is because the impact parameter of m = ±1 scattering, being of order λ, must eventually become much larger than the range of the attractive part of the potential, with the result that the contribution of δ_1 to the heat capacity must be reduced. It follows that at sufficiently low temperatures, the δ_0 contribution, reflecting the hard-core interaction, will dominate the He^3 results as it does the He^4. Consequently the contribution of the second virial coefficient to the He^3 heat capacity will increase with decreasing temperature. This increase occurs at about 0.3K and is shown in Fig. 6. A rise in the He^3 heat capacity has been observed at this temperature.(17)

In a future communication, we shall examine the low-temperature He^3 data, consider the possibility of liquefaction in both isotopes, discuss He^3-He^4 mixtures, and present results for the virial expansion of the magnetic susceptibility of He^3.

ACKNOWLEGEMENTS

We wish to thank R. L. Elgin for providing us with his data prior to publication. We are grateful to M. Bretz, J. G. Dash, and O. E. Vilches for a stimulating series of discussions and to Sir R. E. Peierls for a seminal conversation.

REFERENCES

* Research supported in part by the National Science Foundation.
1. M. Bretz and J. G. Dash, Phys. Rev. Lett. 26, 963 (1971).
2. M. Bretz, J. G. Dash, D. C. Hickernell, E. O. McLean and O. E. Vilches, Phys. Rev. A (to be published).
3. A. D. Novaco and F. J. Milford, Phys. Rev. A5, 783 (1972); D. E. Hagen, A. D. Novaco and F. J. Milford, Proceedings of the Symposium on Adsorption - Desorption Phenomena, Florence, Italy, April 1971 (Academic Press, New York, 1972).
4. C. E. Campbell, J. G. Dash, and M. Schick, Phys. Rev. Lett. 26, 966 (1971).
5. A. Widom and J. B. Sokoloff, Phys. Rev. A5, 475 (1972).
6. A. D. Novaco, J. Low Temp. Phys. 9, 457 (1972).
7. H. W. Chan, A. W. Yanof, F. D. M. Pobell and J. D. Reppy, Proc. 13th Int. Conf. Low Temp. Physics, Boulder, Colo. 1972 (to be published); M. Chester, L. C. Yang, and J. B. Stephens, Phys. Rev. Lett. 29, 211 (1972); J. A. Herb and J. G. Dash, Phys. Rev. Lett. 29, 846 (1972).

8. P. C. Hohenberg, Phys. Rev. 158, 383 (1967).

9. J. M. Kosterlitz and D. J. Thouless, J. Phys. C, 5 L124 (1972).

10. W. A. Steele and E. J. Derderian, Proceedings of the Symposium
    on Adsorbtion – Desorption Phenomena, Florence, Italy, April
    1971 (Academic Press, New York, 1972).

11. K. Huang, "Statistical Mechanics", (J. Wiley and Sons, New York,
    1963) Ch. 14.

12. J. de Boer and A. Michels, Physica 5, 945 (1938).

13. A. Bagchi, Phys. Rev. A3, 1133 (1971).

14. The method of extracting thermodynamic functions from the data
    is to be found in R. L. Elgin and D. L. Goodstein, Proc. 13th
    Int. Conf. Low Temp. Phys., Boulder, Colo. 1972, to be published.

15. The $0.027(Å)^{-2}$ He^3 data is from E. O. McLean, Ph.D. disserta-
    tion, University of Washington (1972), unpublished. The rest
    is from Ref. 1 where they are denoted by coverages x = 0.39
    for He^3 and 0.255, 0.372 for He^4.

16. The data in Figs. 4 and 5 which are in addition to that shown
    in Fig. 3 were taken by the University of Washington group.

17. D. C. Hickernell, E. O. McLean, and O. E. Vilches, Phys. Rev.
    Lett. 28, 780 (1972).

THE ADSORBED HELIUM FILM: TWO DIMENSIONALITY VERSUS REALITY[*]

Anthony D. Novaco

Brookhaven National Laboratory

Upton, New York

1. INTRODUCTION

The lesson of monolayer physics is certainly the usefulness of two-dimensional models in the understanding and interpretation of experimental results.[1] This is especially true for those systems using graphite-like substrates and, in particular, those systems in which helium is the adsorbed species. The two-dimensional nature of adsorption on the graphite-like substrates is due to both the strong interaction of gases with these solids, and the fact that the surfaces of these substrates are unusually clean and nearly ideal.[2,3]

Helium adsorbed upon graphite shows a diversity of two-dimensional behavior.[4] At low densities - and at not too low a temperature - the states of adsorbed helium are quite well described by two-dimensional nonideal classical and/or quantum gas models.[5] At higher densities, there is the superlattice state having a specific heat which behaves as $\exp(-\epsilon/k_B T)$, and showing a "second order" phase transition.[6] For this phase, a two-dimensional lattice gas model gives a good description of the thermal behavior.[7,8] Then at high densities and low temperatures, the thermodynamics indicates a two-dimensional solid with the characteristic T^2 heat capacity.[9]

Yet these systems are only approximately two-dimensional in nature, and at best they are quasi-two-dimensional. That is, the adatoms are essentially confined to a plane, but with some zero-

[*]Work supported by the U. S. Atomic Energy Commission.

point oscillation normal to this plane. An important question is therefore, in what way must the third dimension be incorporated into the calculations, and what is the quantitative effect upon the two-dimensional aspect of the problem?

The discussion will be restricted to the solid phase - that is the high density and low temperature part of the phase diagram - and will ignore any difficulties generated by the lack of long-range order in two dimensions at finite temperatures.[10-13] The philosophy here is simply that the ordered crystal can be a very good starting point for the discussion of a slightly disordered or amorphous solid. The intention is to address two somewhat separate yet quite related questions. First, what are the essential ingredients in a first-principles calculation which can adequately explain the experimental data relating to the ground state of the solid phase? Second, what can be learned about the lattice dynamics of the solid from a simple phenomenological model?

2. GROUND-STATE PARAMETERS

It is quite possible to account for the empirical ground-state properties - at least for areal densities greater than $0.1A^{-2}$ - by a variational calculation in which the helium is assumed to be a quasi-two-dimensional close-packed solid.[14] In this calculation the structure of the monolayer is assumed to be insensitive to the structure of the graphite surface, and the three-dimensional substrate potential $U(\vec{r},z)$ is replaced by the laterally averaged potential

$$\overline{U}(z) = \frac{1}{\Omega} \int_{\Omega} d\vec{r} \; U(\vec{r},z) \tag{1}$$

where $\vec{r}$ is the lateral position vector, z is the normal coordinate, and Ω is the area of the graphite surface unit cell. The helium-helium interaction is described by the Beck potential - semiempirical potential for two helium atoms in vacuum. The helium atoms are assumed to form a simple two-dimensional hexagonal lattice with both in-plane and normal zero-point oscillations. The zero-point oscillations in the plane are adequately described by Gaussian factors in the ground-state wave function, with the width of the Gaussians determined variationally. However, the zero-point oscillations normal to the plane require a more detailed description, one which is dependent upon the shape of the potential $\overline{U}(z)$. Thus, the function M(z) which describes these zero-point oscillations is expanded in an orthonormal set of functions, with the coefficients determined variationally. Finally, the large zero-point oscillations in the plane coupled with the strongly repulsive core of the helium-helium potential necessitates the use of correlation factors in the wave function for an accurate description of the ground state. These correlation factors need depend only upon the lateral separation r at least for the helium-graphite system. The wave function thus takes the form

$$\Psi(\vec{r}_1, z_1 \ldots \vec{r}_N, z_N) = \prod_i \psi_i(\vec{r}_i) M(z_i) \prod_{j < \ell} \prod f(r_{j\ell}) \qquad (2)$$

where each $\psi_i(\vec{r}_i)$ is a Gaussian centered at lattice site $\vec{R}_i$. The functional form used for $f(r)$ in reference (14) is

$$f(r) = {}^{-u(r)[1+P_1 r]}, \quad P_1 = -0.4A^{-1}, u(r) = -\int dr \sqrt{\frac{M\tilde{v}(r)}{h^2}}, \qquad (3)$$

and $\tilde{v}(r)$ is the exponential repulsion of the Beck potential. If the ground-state energy $E = \langle \Psi | H | \Psi \rangle$ is evaluated using the cluster expansion[15] truncated at the two-body term, then

$$E = \sum_i \langle \psi_i(\vec{r}_i) | (-\hbar^2/2M) \nabla_i^2 | \psi_i(\vec{r}_i) \rangle$$

$$+ \sum_i \langle M(z_i) | (-\hbar^2/2M) \frac{\partial^2}{\partial z_i^2} + \bar{U}(z_i) | M(z_i) \rangle \qquad (4)$$

$$+ \langle \psi_i(r_i)\psi_j(r_j) | V_{2D}(r_{ij}) | \psi_i(r_i)\psi_j(r_j) \rangle$$

$$V_{2D}(r) \equiv \frac{\langle M(z) | v(\sqrt{r^2+z^2}) | M(z) \rangle - (\hbar/2M)\nabla^2 \ln f(r)}{\langle \psi_i \psi_j | f^2(r_{ij}) | \psi_i \psi_j \rangle} f^2(r) \qquad (5)$$

where M is just the mass of the helium atom and ∇^2 is the two-dimensional laplacian. In minimizing E with respect to the variational parameters, the dependence of the last term in Equation (4) upon the function M(z) was ignored. Thus the expansion coefficients for M(z) were determined by variation of the second term in Equation (4), and once these coefficients were determined, the first and third terms in Equation (4) were varied with respect to the Gaussian width parameter. In this way, the amplitude of zero-point oscillation normal to the plane is decoupled from the amplitude of the in-plane oscillation.

Once the ground-state energy E is determined as a function of the areal density ρ, the chemical potential, spreading pressure, and compressibility can be found by differentiation.[14] The agreement between theory and experiment is typified by Fig. 1, where the inverse compressibility at 4.2°K[16] is compared to the above

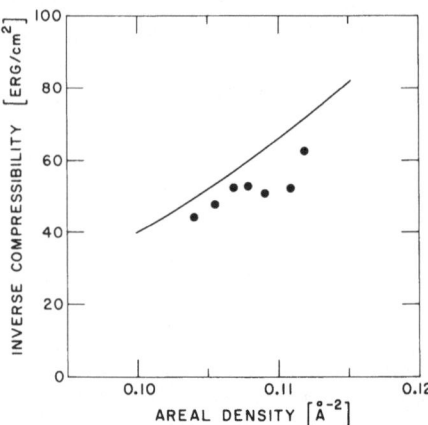

Fig. 1. The inverse compressibility. Solid line is theory at zero°K and the dots are experiment at 4.2°K (reference 16).

calculation. Since the theory and experiment differ by an amount which corresponds to only a percent or two shift in the experimental density, the agreement must be considered more than adequate. The vibrational softening of the potential - the first term in the numerator in Equation (5) has about the same effect as the correlations - the second term in the numerator - and each lowers the inverse compressibility, the chemical potential, and the spreading pressure by about 10 percent from the value found by doing a strictly two-dimensional calculation without correlations.

3. LATTICE DYNAMICS

The calculation described in the previous section does not yield a reasonable phonon spectrum since it is basically an Einstein oscillator model of the solid. However, a standard lattice dynamics calculation[17] for the adsorbed solid will not yield reasonable answers either. It will not do so for the same reason it fails for the bulk helium solid, the zero-point motion of the atoms about their lattice sites is too large.[18] What is done to treat the bulk solid - and presumably what could be done to treat the adsorbed solid - is to combine a Jastrow correlation factor or a t-matrix approach with a correlated Gaussian factor.[19,20] The Jastrow factor or the t-matrix accounts for the short-range correlations and the correlated Gaussian provides a basis for a phonon description of the solid. The correlated Gaussian is the ground state of some fictitious harmonic system whose phonon spectrum is the same as the actual anharmonic solid. However, the parameters which describe this ground state are calculated variationally and approximately correspond to the derivatives of the

effective interatomic potential or the derivatives of the t-matrix
averaged over the zero-point of the atoms rather than simply evaluated at the lattice spacing.[21] Some care must be used in such a
combined approach so that the phonons derived from the ground state
are the same as those which describe the excited states, but this
can be done.[22] The treatment of the adsorbed solid is more complicated than that of bulk because of the failure of a Gaussian
wave function to properly describe the zero-point motion perpendicular to the surface. Thus the proper treatment of phonon modes
polarized in this direction will involve some marriage of correlated Gaussians and single-particle functions like M(z). Such a
first-principles calculation is beyond our purposes here, especially since the theory of the bulk solid is not yet in a totally
satisfactory state.[23] However, the fact that there does exist
a harmonic system which describes the lattice dynamics of the anharmonic solid can be useful in motivating a phenomenological calculation of the lattice dynamics of the adsorbed solid.

The task is to do a harmonic calculation - that is a Born-von
Karman analysis - of the adsorbed solid lattice dynamics using a
phenomenological potential with parameters to be fitted via experiment. Since the hexagon lattice seems a reasonable starting point-
especially in light of the results of Section 2 - we shall treat
the case of a simple two-dimensional hexagonal lattice constrained
in the third dimension by an external harmonic well which, of course,
mimics the substrate potential. Because the basic characteristic
of the adsorbed solid is that it exists only at finite pressure,
the dominant part of the actual potential is the repulsive part.
This is exponential in form - for the Beck potential - as are its
derivatives. Therefore, as a phenomenological interatomic potential we will use an exponential repulsion with two parameters to
be experimentally determined. The nearest neighbor Born-von Karman
analysis requires the first and second derivatives of this potential evaluated at the lattice spacing corresponding to the solid
areal density ρ. If these are denoted by v' and v'' respectively,
these can be conveniently expressed as

$$v' = - \frac{v''}{B} \left[\rho_0 \cos \left(\frac{\pi}{6}\right) \right]^{-\frac{1}{2}} \quad , \tag{6}$$

$$v'' = M\omega_0^2 \, e^{B(1 - \sqrt{\rho_0/\rho})} \quad , \tag{7}$$

where ρ_0 is a conveniently chosen density, M is the mass of the
helium atom, and the constants ω_0 and B are determined by fitting
the experimental Debye temperature as a function of density. The
characteristic frequency ω_p of the external harmonic potential was
set at 60°K, which is some compromise between theoretical[24] and

experimental[25] values for the excitation energy of the lowest
single-particle state exhibiting excited oscillations normal to the
surface.

The phonon frequencies $\omega(\vec{q})$ for a two-dimensional phonon with
wave vector $\vec{q}$ is given by the square roots of the eigenvalues of
the dynamical matrix which is defined by

$$D^{\alpha\beta}(\vec{q}) \equiv \omega_p^2 \, \delta^{\alpha\beta}\delta^{\alpha z} + \frac{1}{M}\sum_{j=1}^{6} \, \Phi^{\alpha\beta}(R_j)\left[1 - \cos(\vec{q}\cdot\vec{R}_j)\right] \qquad (8)$$

$$\Phi^{\alpha\beta}(\vec{R}_j) = v''(R_j)\left[\frac{\partial\eta}{\partial\alpha}\frac{\partial\eta}{\partial\beta} - \frac{[\rho_o \cos(\frac{\pi}{6})]^{-\frac{1}{2}}}{B}\frac{\partial^2\eta}{\partial\alpha\partial\beta}\right] \qquad (9)$$

where α and β denote x, y, or z and $\eta^2 \equiv x^2 + y^2 + z^2$. Because the
z = 0 plane is a plane of mirror symmetry, the dynamical matrix is
block diagonal and has the form

$$\overset{\leftrightarrow}{D} = \begin{pmatrix} D^{xx} & D^{xy} & 0 \\ D^{yx} & D^{yy} & 0 \\ 0 & 0 & D^{zz} \end{pmatrix} \qquad (10)$$

Thus there are three phonon modes, one polarized normal to the
surface and two whose polarization vectors lie in the plane. The
latter two have frequencies that tend to zero linearly as q → 0 –
"acoustic" phonons – while the spectrum of the former – "optical"
phonon – tends to ω_p in the same limit.

In the long-wavelength limit, the two acoustic modes are iso-
tropic with the longitudinal and transverse sound velocities given
by

$$C_L^2 = \frac{\sqrt{3}}{4}\frac{\omega_o^2}{\rho}\,e^{B(1 - \sqrt{\rho_o/\rho})}\left[3 - \frac{1}{B}\sqrt{\rho/\rho_o}\right] \qquad , \qquad (11)$$

$$\frac{C_L^2}{C_T^2} = \frac{3B - \sqrt{\rho/\rho_o}}{B - 3\sqrt{\rho/\rho_o}} \qquad . \qquad (12)$$

The Debye temperature can now be calculated from Equations (11) and (12) since Θ_D is just given by

$$\Theta_D = \frac{2\pi\hbar}{k_B}\left[\frac{2\rho}{\pi}\right]^{\frac{1}{2}}\left[\frac{1}{c_L^2} + \frac{1}{c_T^2}\right]^{-\frac{1}{2}} \qquad (13)$$

For values of B greater than 10,

$$\ln \Theta_D = \ln \Theta_D(\rho_o) + \frac{B}{2}\left[1 - \sqrt{\rho_o/\rho}\right] \qquad (14)$$

with an error of about 1% or less.

Figure 2 shows a least squares fit of Equation (14) to the experimental data.[9] The fitted value of B is 16.0 and the fitted value of $\hbar\omega_o/k_B$ is 5.8°K. The sound velocity can now be calculated. For $\rho = 0.112$ A^{-2} – close to a completed monolayer – the theoretical values of c_L and c_T are 883 m/sec and 455 m/sec. There is no direct measurement of these, but there are empirical values[16] based upon an elastic theory analysis of the Debye temperature and compressibility at $\rho = 0.11$ A^{-2}. These values are 945 m/sec for c_L and 420 m/sec for c_T. Given the uncertainty in both calculations, the agreement is quite adequate. The phonon spectrum along the high symmetry direction in the Brillouin zone is shown in Fig.3. Along these directions, the phonon modes are either purely longitudinal or purely transverse.

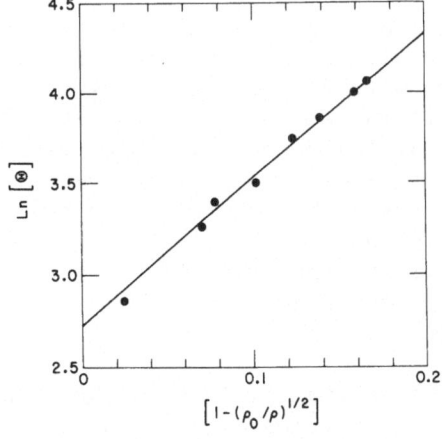

Fig. 2. Least squares fit of Equation (14) – solid line – to the experimental values – dots – of Reference (9).

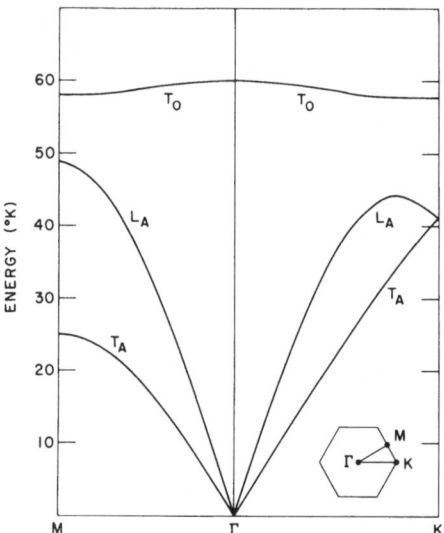

Fig. 3. Theoretical phonon spectrum for the quasi-two-dimensional helium solid with an areal density of 0.112 A^{-2}. The longitudinal and transverse "acoustic" branches are polarized in the plane, while the "optical" branch is polarized normal to the plane. The insert shows the Brillouin zone.

4.   CONCLUSIONS

The most notable aspect of the phonon spectrum is the narrow width of the "optical" branch. This width is a measure of the coupling between the amplitudes of in-plane and normal-to-the-plane zero-point motion. The fact that the energy of this branch is lower at the zone boundary than at zone center is a direct consequence of the repulsive interaction between the helium atoms, and this shift increases with density. The magnitude of this shift relative to the shift in the lower curves gives an estimate of the effect of the third dimension upon the compressibility, and it can be seen to have a very small effect. The narrowness of the upper branch is due to the large value of ω_p compared to ω_0. This ratio is about 10 for the graphite-like substrates, but can be much less for others. As ω_p is lowered, the effect of the upper branch upon the compressibility - and even upon the stability of the film - becomes more important.

Besides giving a general picture of the phonon modes in the adsorbed solid, the calculation of Section 3 can be useful in analyzing experimental data such as inelastic neutron scattering and the temperature dependence of the Debye temperature. Furthermore, with slightly more work, the phenomenological model of Section 3 can be very instructive in the examination of the effect of third dimension upon the "melting" peaks observed in the heat capacity.[9] This would involve a full self-consistent phonon calculation using the phenomenological potential of Section 3 with ω_0 and B recalculated from the correlated Gaussian ground state. As the temperature is increased, the "optical" zone boundary phonon should soften. If this effect is large enough, then this could have an important effect upon the nature of the "melting" transition.[26] Finally, the anharmonic nature of the substrate-adatom system should increase the coupling between the "optical" and "acoustic" modes and thus further amplify this effect.[27]

ACKNOWLEDGMENT

The author would like to thank E. Eisenriegler for fruitful discussions.

REFERENCES

1. J. G. Dash, J. Low Temp. Phys. $\underline{3}$, 301 (1970).
2. A. D. Crowell and R. B. Steele, J. Chem. Phys. $\underline{34}$, 1347 (1961).
3. S. Ross and J. P. Olivier, "On Physical Adsorption," (Interscience, New York, 1964) Chapter VII, p. 186.
4. M. Bretz, J. G. Dash, D. C. Hickernell, E. O. McLean, and O. E. Vilches, Phys. Rev. A (to be published).
5. W. A. Steele and E. J. Derderian, in "Adsorption-Desorption Phenomena," ed. F. Ricca (Academic Press, London, 1972). Also R. L. Siddon and M. Schick, this conference.
6. M. Bretz and J. G. Dash, Phys. Rev. Letters $\underline{27}$, 647 (1971).
7. C. E. Campbell and M. Schick, Phys. Rev. A $\underline{5}$, 1919 (1972).
8. M. Schick and R. L. Siddon, Phys. Rev. A $\underline{8}$, (1973).
9. M. Bretz, G. B. Huff, and J. G. Dash, Phys. Rev. Letters $\underline{28}$, 729 (1972).
10. N. D. Mermin, Phys. Rev. $\underline{176}$, 250 (1968).
11. H.-J. Mikeska and H. Schmidt, J. Low Temp.Phys. $\underline{2}$, 371 (1970).
12. B. Jancovici, Phys. Rev. Letters $\underline{19}$, 20 (1967).
13. J. M. Kosterlitz and D. J. Thouless, J. Phys. C $\underline{6}$, 1181 (1973).
14. A. D. Novaco (to be published).
15. A. D. Novaco, Phys. Rev. A $\underline{7}$, 1653 (1973).
16. G. A. Stewart, S. Siegel, and D. L. Goodstein, Proc. 13th Int. Conf. Low Temp. Phys. Boulder, Colorado, 1972 (1973).
17. M. Born and K. Huang, "Dynamical Theory of Crystal Lattices (Clarendon Press, Oxford, 1954).
18. R. A. Guyer, in "Solid State Physics," edited by F. Seitz, D. Turnbull, and H. Ehrenreich (Academic Press, New York, 1969), Vol. 23.
19. N. R. Werthamer, Am. J. Phys. $\underline{37}$, 763 (1969).
20. H. Horner, Z. Phys. $\underline{205}$, 72 (1967).

21. M. L. Klein and G. K. Horton, J. Low Temp. Phys. $\underline{9}$, 151
     (1972).
22. N. R. Werthamer, Phys. Rev. A $\underline{7}$, 254 (1973).
23. S. B. Trickey, N. M. Witriol, and G. L. Morley, Phys. Rev.
     A $\underline{7}$, 1662 (1973).
24. D. E. Hagen, A. D. Novaco, and F. J. Milford, in "Adsorption-
     Desorption Phenomena," edited by F. Ricca (Academic Press,
     London, 1972).
25. R. L. Elgin, Thesis (Cal. Inst. Tech. 1973) unpublished, and
     R. L. Elgin and D. L. Goodstein, Proc. 13th Int. Conf. Low
     Temp. Phys.,Boulder, Colorado, 1972 (1973).
26. Some evidence of the importance of this kind of effect has
     been reported in Reference 25.
27. The "optical" branch will, in fact, not be polarized purely
     in the direction normal to the surface since the dynamical
     matrix will, in general, no longer be block diagonal.

PHYSISORBED HELIUM MONOLAYERS AND BULK HELIUM SURFACES

Chia-Wei Woo

Department of Physics, Northwestern University

Evanston, Illinois 60201

As might be gathered from the title of the talk, the message that I wish to bring to this Symposium is that the problems concerning physisorbed helium monolayers and bulk helium surfaces are in many ways one and the same.

On account of the ease with which an experimenter can control the adsorption coverage, helium monolayers can be prepared over a much wider range of (areal) densities than He^3 dissolved in bulk He^4. As one increases the coverage, a physisorbed He monolayer is expected to traverse successive phases of quantum gas, quantum liquid, and quantum crystal. All this in addition to its appearance as a real-life two-dimensional rig! For these reasons, I believed, along with undoubtedly many others attending the last Stevens Symposium[1], that in three years He monolayers would reign as one of the hottest systems in low temperature physics. Thanks in part to grafoil, they have become precisely that. However, I have been somewhat disappointed at one aspect of the way in which the field has developed: in terms of <u>active</u> participation, its growth has not been spectacular. In talking with Academician Khalatnikov recently, I discovered that the subject which has provided us with so much fun and excitement here has not received much attention at all in the Soviet Union. It is all the more surprising since over the years so much quality research has been performed in the Soviet Union on He films and bulk He surfaces. Recently in this country excellent work has been carried out at e.g. Ohio State and Argonne on the surface tension, ripplon spectrum, and ion mobility in bulk He liquids and solutions. If it can be convincingly advocated that in many facets we are really dealing with the same problem, it will surely strengthen communication lines within the low temperature field and increase

our basic understanding of the underlying physics. With this
purpose in mind, I wish to analyze a very simple problem today
and look at if from two angles. I shall discuss the behavior of
a He^3 atom physisorbed on bulk He^4 surface.

PHYSISORPTION OF He^3 ON A He^4 SURFACE

Physisorption describes a process in which a neutral atom
or molecule by virtue of van der Waals attraction becomes bound
to the surface of a condensed substrate. The crudest model that
can be constructed for the substrate is a static, semi-infinite
single-crystalline lattice bounded by a sharp surface. The field
experience by an atom in the neighborhood of such a surface can
be obtained by summing over its interaction with each individual
substrate atom. Let the position of the adatom be denoted by $\vec{r}$
and those of the substrate atoms by $\vec{R}_\alpha$. The adsorption potential
is given by:

$$V(\vec{r}) = \sum_\alpha v(|\vec{r}-\vec{R}_\alpha|). \tag{1}$$

Plotted with respect to the normal distance z of the adatom from
the surface, $V(\vec{r})$ appears as a series of wells, one for each lateral position (x,y). These curves merge at large z, and rise
sharply as z→0. On each curve, the size of the repulsive region
and the depth of the well depend strongly on the proximity of
(x,y) to an adsorption site. To determine whether physisorption
takes place and to study the motion of the adatom, one must solve
the quantum mechanical problem defined by the single-particle
Hamiltonian

$$H = \frac{-\hbar^2}{2m} \nabla^2 + V(\vec{r}). \tag{2}$$

Much effort on the part of Milford, Novaco, and some of my co-
workers has been devoted to the solution of the one-body Schrödinger Equation. Had the potential been separable into normal and
lateral parts, the problem would have reduced to a simple exer-
cise. There would be bound and scattering states in the direction normal to the surface, and band structure in the lateral
plane consistent with the periodicity of the substrate lattice.
As it actually happens, there is much mixing of states, and consequently even for such a crude model the solution requires
painstaking labor. Furthermore, at any finite coverage one is
dealt a many-body problem. Starting with these very unwieldy
single-particle functions, the many-body calculation becomes ex-
ceedingly complicated. While the experimenter busies over the
preparation of a clean, homogeneous adsorption surface, the theorist finds himself in the embarassing position of not knowing
what to do even if he is handed a perfect crystalline surface!
For him, even a perfect crystalline substrate is far from ideal.

What then constitutes an ideal surface physisorption? Clearly, in the view of a theorist it is best to have as little lateral variation on the surface as possible. This is not only for the convenience of microscopic calculations, but also to preserve in a maximal way the two-dimensional nature of the problem. To cite a concrete example, let us consider the formation of a two-dimensional quantum crystal. If the substrate surface is microscopically smooth, to the extent that adsorption sites become unidentifiable, one can be certain that whatever two-dimensional lattice formed in the adsorbed layer will be self-bound: A Debye solid, whose properties are well-known. On the other hand, as long as the periodicity of the substrate is discernible, one finds it impossible to decouple the adsorbed from the adsorbing. In making measurements on the monolayer, then, it is not clear whether one is studying the properties of the adsorbed layer, or those of the substrate surface using the adatoms as mere probes. This results in many complications starting with the phenomenon of misregistry.

Solid fcc argon served in several calculations as an idealization of argon-plated copper substrates. If the (001) face is taken to be the substrate surface, the lateral variations can be quite significant. One finds narrow bands and wide gaps, and nearly localized adsorption. When the (111) face is used, a close-packed hexagonal (triangular) lattice appears, yielding much broader adsorption bands and much narrower gaps. The adatoms become relatively mobile. Over the last two year, Dash and coworkers have taken great strides toward reaching the ideal. On grafoil, the high mobility of the adatoms is evidenced by specific heat Nk_B at high temperatures and low coverages. Yet it must be concluded that the ultimate ideal can be achieved only by adopting a liquid substrate. There the lateral variation vanishes altogether.

One obvious condition on using a liquid substrate is that its freezing point must be low compared to the heat of desorption (in units of k_B). This rules out e.g. helium adsorbed on liquid argon. In fact we are left with but one choice: liquid helium itself. If He^4 is chosen to serve as the substrate, He^3 will make a perfectly convenient choice of adatom.

Imagine a full layer of He^3 placed on the surface of bulk He^4. In reality the distribution of He^4 does not end abruptly; it fades off in the direction normal to the interface, seeping weakly through the He^3 layer. The He^3 atoms generally reside on the surface; only occasionally will they wander a bit into the interior. For idealization, we neglect for the moment the overlap of He^3 and He^4 wave functions. We cut off the He^4 liquid at a sharp surface and consider its influence on the motion of a single He^3 atom.

Following the prescription stated in Eq. (1), the adsorption potential is first obtained via summation. Let v be represented by a Lennard-Jones 6-12 potential:

$$v(|\vec{r}-\vec{R}|) = 4\epsilon[(\frac{\sigma}{|\vec{r}-\vec{R}|})^{12} - (\frac{\sigma}{|\vec{r}-\vec{R}|})^6] \qquad (3)$$

$V(\vec{r})$ is then calculated by integrating $v(|\vec{r}-\vec{R}|)$ with respect to $\vec{R}$ over an infinite half-space. As a result,

$$V(\vec{r}) \equiv V(z) = 4\pi\rho_0\epsilon\sigma^3[\frac{1}{45}(\frac{\sigma}{z})^9 - \frac{1}{6}(\frac{\sigma}{z})^3] , \qquad (4)$$

where ρ_0 is the equilibrium density of liquid He^4. $V(z)$ has a minimum of -8.2°K at $z = z_{min} = 2.19$Å. For a quick estimate, one might expand the potential about z_{min}. A harmonic coefficient (spring constant) of 46°KÅ^{-2} results. For a He^3 atom, this represents a quantum of 27°K, or a zero-point energy of $1/2 \hbar\omega = 13.5$°K. The ground state at 13.5°K-8.2°K = 5.3°K, is thus unbound. However, it is clear from the form of the potential curve that a harmonic approximation is most unreliable. A narrow Gaussian wave function centered at z_{min} loses a large chunk of the attractive region, in addition to gaining much unneeded kinetic energy when it dies off abruptly in the region $z>z_{min}$. An exact calculation may give rise to one or more bound states.

Such a calculation was made by M.D. Miller. The resulting wave function is appropriately skew, decaying slowly at large z. He found just one bound state at -0.79°K.

In the realistic case, the binding energy is expected to be larger for the following reasons. First of all, the He^4 surface is not completely sharp. As will be seen in the next section, it spreads over a thickness of 2-3Å. Therefore the He^3 wave function need not drop off too abruptly in the region $z<z_{min}$. This means a lower kinetic energy. This effect is however expected to be small. Next, the smearing out of the He^4 atoms as implied by the integration procedure grossly under-estimates the potential depth. In a calculation such as the present one, the He^4 atoms do not take part in the motion. There is no way to account for their "graininess", which nevertheless has two important effects. Think momentarily of the He^4 substrate as made up of a simple cubic lattice. Directly above an adsorption site it is easy to find a location for a He^3 to sit so as to be $\sqrt[6]{2}$ σ or 2.89Å away from each of four He^4 atoms. Lattice summation tells us that at that position $V(z)<-4\epsilon \equiv -40.88$°K. Imagine now the smearing out of the cubic lattice. $V(z)$ will rise as each of the four He^4 atoms dissolves around the potentail minimum of He^3. The reduction of the well-depth to 8.2°K is thus seen to be an artifact of our very crude model. Also, the graininess of the liquid permits the <u>absorption</u> of He^3 atoms. In microscopic terms, the infinite <u>repulsive</u> wall sampled by He^3 in the present model is also artificial. The potential should rise to a finite value and thereafter turn constant in the interior of the liquid substrate.

SURFACE STRUCTURE OF LIQUID He^4

Let us discuss briefly the structure of our substrate surface.[2] The information we need includes the thickness of the surface layer and the energy spectra of surface excitations.

Place N He^4 atoms in a box of volume $2L^3$. The coordinates x and y extend from 0 to L, while z extends from -L to L. Let N and L approach infinity such that the <u>average</u> density $N/2L^3$ equals $1/2 \, \rho_0$. At zero temperature, the atoms will choose to congregate in the lower half-space $z \leq 0$. The density throughout the liquid will be ρ_0, with the exception that the topmost layers will dilute and spread into the upper half-space. It is the structure of the liquid in the vicinity of the vacuum-liquid interface that we now wish to investigate.

Take a trial wave function of the form

$$\psi_4 = \prod_{1 \leq i < j \leq N} e^{1/2 \, u(r_{ij})} \prod_{\ell=1}^{N} \varphi_0(z_\ell) \, . \tag{5}$$

At large distances inside the free surface, $u(r)$ must approach its optimum form for the ground state of bulk He^4. Approximately, we have from previous work:

$$u(r) = - \left(\frac{a\sigma}{r}\right)^5 , \tag{6}$$

with $a = 1.17$. We are left then with one variational function $\varphi_0(z)$. The energy expectation value can be expressed as a functional of $u(r)$, $\varphi_0(z)$, and the one- and two-particle distribution functions $n(\vec{r}_1)$ and $n(\vec{r}_1, \vec{r}_2)$ defined by the general formula:

$$n(r_1, r_2, \ldots, r_n) = N(N-1) \ldots (N-n+1) \frac{\int \psi_4^2 \, dr_{n+1}, \ldots, N}{\int \psi_4^2 \, d\vec{r}_{1, \ldots, N}} . \tag{7}$$

It is a simple matter to derive an heirarchy of integro-differential equations relating the distribution functions $u(r)$ and $\varphi_0(z)$. The first of the heirarchy determines $n(\vec{r}_1)$, or $n(z_1)$:

$$\frac{dn(z_1)}{dz_1} = n(z_1) \frac{d\ell n \varphi_0^2(z_1)}{dz_1} + \int d\vec{r}_2 \frac{z_{12}}{r_{12}} \frac{du(r_{12})}{dr_{12}} n(\vec{r}_1, \vec{r}_2) . \tag{8}$$

Instead of obtaining $n(\vec{r}_1, \vec{r}_2)$ from equations involving higher-order distribution functions, we take the following model:

$$n(\vec{r}_1, \vec{r}_2) \approx n(z_1)n(z_2)g_b(r_{12}|\rho = \sqrt{n(z_1)n(z_2)}\,),\quad (9)$$

where $g_b(r|\rho)$ is the bulk radial distribution function at hypothetical densities $\rho \leq \rho_0$. The result of solving Eqs. (8) and (9) and minimizing the energy expectation value can be summarized as follows. (i) The surface structure is approximately described by the density variation

$$n(z) \approx \frac{\rho_0}{1 + \exp(\beta z)} \qquad (10)$$

with $\beta \approx 1.7 \overset{\circ}{A}{}^{-1}$. The surface layer is therefore about 2-3 Å in thickness. (ii) The surface energy per unit area, which at zero temperature corresponds to surface tension, is 0.36 dyne/cm. (Experimental value: 0.37-0.38 dyne/cm.)

So much about the ground state of bulk He[4] with a free surface.

In first approximation, the surface excitations can be written after the fashion of Feynman in his description of bulk phonons. We have:

$$|\vec{K}, \alpha\rangle = \sum_{\ell=1}^{N} \frac{\omega_\alpha(z_\ell)}{\omega_0(z_\ell)} e^{i\vec{K}\cdot\vec{\rho}_\ell} \psi_4 , \qquad (11)$$

where $\vec{K}$ and $\vec{\rho}$ are two-dimemsional vectors in respectively the momentum and coordinate space. Jackson[3] proposed such a description of "longitudinal surfons" ($\alpha=0$) and "transverse surfons" ($\alpha>0$) for physisorbed He monolayers in the extreme mobile limit. In the present case, the states given in Eq. (11) represent quantized capillary waves, or ripplons. It should be noted, however, that in a calculation of the ripplon spectrum, appropriate linear combinations of these states must be used.

The Ohio State group has pioneered current efforts in studying surface excitations. Saam[4] suggested recently that neutron scattering techniques can be employed to probe the ripplon spectrum. M.D. Miller has computed the spectra corresponding to the state functions given in Eq. (11). They will be discussed elsewhere. For the purpose of the present talk, it suffices to note that these surface states play an important role in the formation of physisorbed He[3] layers. This will become clear in the next section.

A word about the model Eq. (9). The same model has recently been applied to study the structure of electron bubbles in liquid He[4] under pressure, and yielded results in spectacular agreement with experiment[4]. Calculations using the same model for studying electrons adsorbed in and on bulk He surfaces are under way. We expect soon to apply it to investigate vortex line structures. It appears that such a model will serve adequately in the role of unifying a number of problems involving density variations.

SURFACE STATES OF He^3 IN DILUTE He^3-He^4 SOLUTIONS

The experiments by Esel'son et. al.[2] first indicated that the presence of He^3 impurity in superfluid He^4 lowers the surface tension, and the extent of lowering exceeds what one might expect from the usual rule of additivity. To intrepret these results, Andreev[2] suggested that there exist surface states of He^3 with energies below that in the bluk. The single He^3 quasiparticle spectrum was taken to be of the form

$$\epsilon_s(k) = -\mu_3 - \epsilon_0 + \frac{\hbar^2 k^2}{2m_s^*} , \qquad (12)$$

where μ_3 denotes the chemical potential of He^3 in bulk solutions, and ϵ_0 represents the extra binding energy that drives He^3 atoms toward the surface layer. Experiments[2] determined μ_3 at 2.79°K and ϵ_0 at 1.7-1.95°K. The binding energy is thus about 4.6°K. Andreev's model has found further support in experiments[2] dealing with vortex-ring creation by negative ions in dilute He^3-He^4 solutions and third sound measurements in He^3-He^4 films.

A moment's reflection will convince us that what Andreev's model describes is precisely the physisorption of a He^3 on He^4 surface. The lateral translational invariance in this problem results in mobile adsorption and permits the definition of an effective mass m*. Zinov'eva and Boldarev[2] determined m_s^* at (0.9+0.1) times the effective mass of He^3 in the bulk, or (2.1+0.2) times the bare mass m_3. Guo et al.[2] found (2.07+0.1)m_3. We interpret these results to mean that there exists important Landau[3] dressing by adsorbed He^3 on one another in the fully degenerate region where the experiment was carried out.

Earlier in this talk, I mentioned that a crude estimate gave a binding energy of 0.79°K. For the reasons outlined there, we expect it to be much smaller than the experimental value. Let me now present a totally different approach toward solving the problem. He^4 coordinates will now take part in the formulation of the problem: they will no longer act as mere parameters.

In a bulk liquid with a free surface we replace the Nth He^4 atom by a He^3 atom. The Hamiltonian changes by a term

$$\left(\frac{\hbar^2}{2m_4} - \frac{\hbar^2}{2m_3}\right)\nabla_N^2 .$$

By incorporating into the trial wave function a correction factor $\zeta(z_N)$, thus:

$$\psi = \psi_4 \zeta(z_N) , \qquad (13)$$

the energy expectation value picks up dependences on two new functions: $\zeta(z_N)$, and the He^3 density defined by

$$n_3(z_N) = N \frac{\int \psi^2 d\vec{r}_{1,\ldots,N-1}}{\int \psi^2 d\vec{r}_{1,\ldots,N}} = \frac{N}{L^2} \frac{n(z_N)\zeta^2(z_N)}{\int n(z_N)\zeta^2(z_N)dz_N} . \qquad (14)$$

The variational calculation that follows yields the conclusions given below. (i) $n_3(z)$ finds itself centered on ghe Gibbs' surface $z=0$. (ii) $n_3(z)$ is nearly symmetrical. Writing it in the Gaussian form

$$\exp\left[-\left(\frac{z}{\lambda}\right)^2\right] ,$$

we find $\lambda=2.5\text{Å}$. (iii) The extra binding energy on the surface is $\epsilon_0=1.6°K$. When added to μ_3, a binding energy of $4.4°K$ results, in good agreement with experiment. (iv) There is only one bound state in the motion normal to the adsorbing surface.

If and when a more detailed calculation is carried out, one should be able to map out the single-particle adsorption potential $V(z)$ mentioned earlier. This will then complete contacts between the two approaches.

Let me conclude with a few words about the formation of a physisorbed He^3 layer. One question which must be answered is whether the presence of He^3 near the surface will encourage other He^3 atoms to move to the surface. This is equivalent to asking whether the effective interaction between He^3 atoms is on the average repulsive or attractive. The interaction includes two parts: direct He-He interaction and interactions mediated by the substrate. The latter is generated by the exchange of surface and bulk excitations between two He^3 atoms. Such forces are necessarily attractive. While it is difficult to calculate the effective interaction between atoms adsorbed on crystalline sub-strates - even if one employs a continuum approximation for the crystal, excitations in He^4 can be described by relatively simple wave functions; consequently I feel quite optimistic about the prospect of a successful calculation. A second question concerns the effective mass. The self-energy related to interaction with the substrate must first be obtained. This calculation is not very different from the effective interaction calculation. Next, using the effective interaction in Landau's Fermi liquid theory, one can evaluate the quasiparticle mass m*, and subsequently the Fermi energy at each coverage. A monolayer becomes complete when the Fermi energy and the effective interaction energy combine to cancel the surface term $-\varepsilon_o$; for at that moment it is no longer energetically favorable for the next He^3 to enter the monolayer.

I wish to acknowledge collaboration and helpful discussions with Professor Yu Ming Shih and Mr. Michael D. Miller. This work was supported in part by the National Science Foundation through Grant No. GP-29130, and by the Alfred P. Sloan Foundation through a Sloan Fellowship.

References

1. Proceedings of the Symposium on Thin Helium Films, J. Low Temp. Phys. 3, 197-337 (1970).
2. Material on surface-states calculations is taken from Y.M. Shih and C.-W. Woo, Phys. Rev. Letters 30, 478(1973). Most of the papers mentioned here are included in the list of references there.
3. H.W. Jackson, Phys. Rev. 180, 184 (1969).
4. Preprints and papers to be published.

HARD SPHERE BOSONS IN A CHANNEL[*]

K. S. Liu[+][1], M. H. Kalos[2] and G. V. Chester[3]

1. Cornell and New York University
2. Courant Institute of Mathematical Sciences
   New York University
   New York, NY 10012
3. Laboratory of Atomic and Solid State Physics
   Cornell University
   Ithaca, NY 14850

INTRODUCTION

We have been engaged for some time in a program of computational investigations of the properties of boson systems in their ground states. It seemed timely to consider the simplest model of a film-like system and we present here some results of our work on these models.

The model we consider is a system of N hard sphere bosons confined between a pair of parallel walls which are represented by infinitely repulsive potentials. In directions parallel to the walls, the wave function is assumed periodic. This is very likely the simplest way of restricting a bulk ensemble so as to resemble a film.

We use two walls for two important reasons: 1) We expect that the structure of the fluid near any surface exhibiting strong repulsive forces would be dominated by those forces, just as the structure of the bulk

[*]Supported by the A.E.C. under Contract No. At(11-1)-3077 and Grant No. GH-36457 with the National Science Foundation.

[+]Present mailing address, Courant Institute of Mathematical Sciences, New York University, New York, NY 10012

system is dominated by the repulsive part of the interatomic forces.   2) We wished to avoid at first the significant complication for work of this kind of dealing
with a free surface. Once the system is bounded either
by walls or more particles, the attractive part of the
potential plays only a small role in determining the
structure. This has been shown most convincingly for
bulk systems by the recent numerical work of Kalos,
Levesque, and Verlet[1].

VARIATIONAL TREATMENT

The first numerical treatment consisted of a fairly
thorough survey of the problem by variational means[2].
Suppose the wall are normal to the z direction. We take
a trial wavefunction of the form

$$\psi = \psi_B(r_1 \ldots r_N) \prod_{i=1}^{N} h(z_i)$$

with ψ_B of the Bijl-Jastrow form

$$\psi_B(r_1 \ldots r_N) = \prod_{i<j} f(r_{ij}) \; ;$$

we use for f that function which Hansen, Levesque and
Schiff[3] introduced for the variational treatment of the
bulk hard core fluid:

$$f(r) = \tanh \left[(r/\sigma)^m - 1 \right]]/b^m$$

where σ is the hard core radius. We also adopted the
values of the parameters m, b which they found optimum
for the ground state. The effect of walls at $z = \pm L_z/2$
is introduced by means of the function h which is given
the functional form.

$$h(z) = \tanh \left[(z + L_z/2 - z)/dL_z \right]^q$$

The energy expectation of the system is readily calculated by Monte Carlo quadrature; we sought the vales of
the parameters q and d which gave the least energy.
Calculations were carried out for the density $\rho\sigma^3 = 0.2$
for N from 16 to 256 to investigate possible size

effects in the z and (x,y) directions. The only significant changes are associated with variation of the film thickness. A more or less typical result is shown in Figure 1 which gives the density profile for a 64 body system for a wall separation of L_z = 6.84σ. This shows clearly the presence of well-developed layers near the two walls.

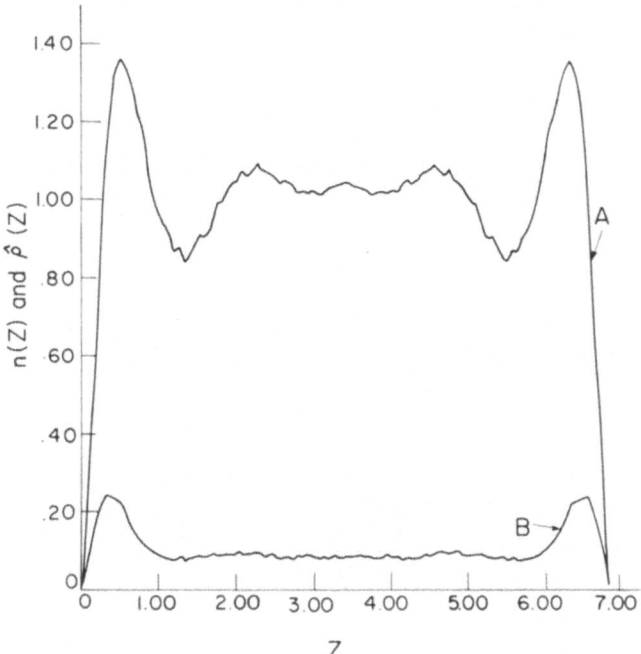

Fig. 1: Curve A, the variational result of density function; Curve B, the variational result of $\hat{\rho}(z)$.

We interpret this structure as analogous to that seen in the radial distribution function near a repulsive core. The rather smooth behavior at the center of the channel appears to confirm our belief that the structure near each wall would be uninfluenced by the presence of the other wall. Figure 1 also gives numerical results for the square of the order parameter, i.e.

$$\hat{\rho}(z) = \lim_{\substack{z=z' \\ |\vec{r}-\vec{r}'|\to\infty}} \frac{1}{n} \frac{\int \sum_1 \delta(\vec{r}_1-\vec{r})\delta(\vec{r}_1'-\vec{r}')\psi(\vec{r}_1,..,\vec{r}_1,..,\vec{r}_N)\psi(\vec{r}_1,..,\vec{r}_1',..,\vec{r}_N)d^{3N}r}{\langle\psi|\psi\rangle}$$

which demonstrates similar structure. Note that the Har-
tree theory would show no such structure.

MORE ACCURATE TREATMENT

We have adapted the computer program developed by
Kalos and Verlet to this wall problem. This method
can in principle carry out an exact numerical integration of the Schrodinger equation by Monte Carlo methods
Figure 2 shows the density profile as calculated in this
way compared with the variational results.

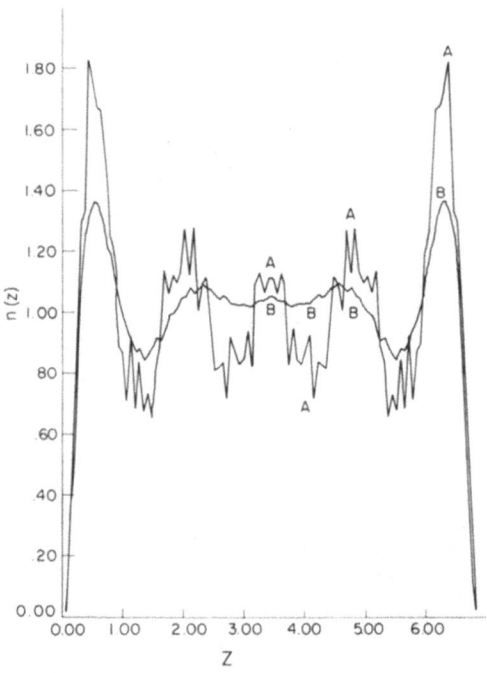

Fig. 2: Curve A, density function from the exact method;
Curve B, that from a variational calculation.

Observe that the new results have greater statistical
error than the variational results. Most significantly,
they show much greater structure across the entire
film. This is similar to but much stronger than the
corresponding effects in bulk where application of exact
methods leads to sharper peaks in the radial distribution function. Incidentally the position of the peaks
relative to the walls follows the positions of peaks in
the bulk r.d.f.. The calculations show very clearly

that a strong arrangement of layers establishes itself
in the gap between two walls without assistance of any
binding effect of the walls. Figure 3 gives a calculation of $\rho(z)$ which is intermediate numerically between
the exact and variational treatment. The new result
seems to follow the more structured density profile.

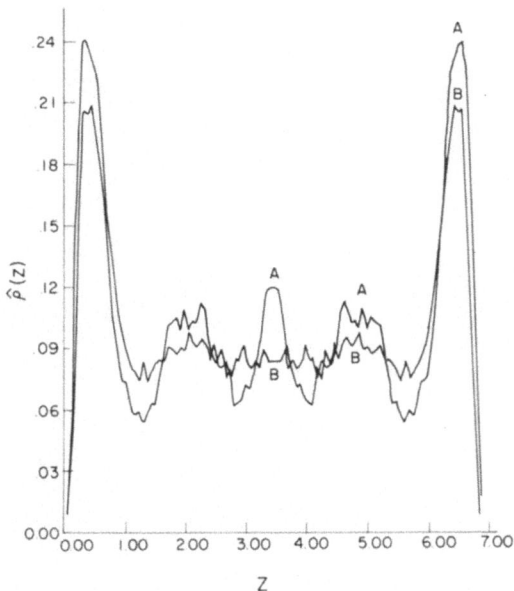

Fig. 3: $\rho(z)$, A from the exact method, and B from a
variational calcualtion.

CONCLUSIONS

Our results show that a simple model of a boson
film exhibits rather significant structure effects
near repulsive walls. Indeed the structure is such
that our original expectation in choosing the model -
that effects at the two walls would be decoupled - may
not be borne out. We plan to examine the effect of
small changes in wall separation to see if the structure
is a result of a kind of constructive interference. It
will be necessary to treat the free surface problem.
We expect to do this next, introducing rather simple
attractive potentials to bind the system.

REFERENCES

1. M. H. Kalos, D. Levesque, and L. Verlet, to be published.

2. K. S. Liu, M. H. Kalos, and G. V. Chester, to be published.

3. J.-P. Hansen, D. Levesque and D. Schiff, Phys. Rev. $\underline{A3}$, 776 (1971)

NUCLEAR MAGNETIC RESONANCE IN ADSORBED HELIUM

D. F. Brewer, D. J. Creswell, Y. Goto,

M. G. Richards, J. Rolt, A. L. Thomson

University of Sussex, Falmer, Brighton, England

INTRODUCTION

There is at present comparatively little information available on the nuclear magnetic properties of adsorbed He^3. Most of the experiments have been done on heterogeneous adsorbing surfaces such as Zeolite[1] or Vycor porous glass[2], but recent work has been initiated with other adsorbents, including the more regular surface of Grafoil[3], as described elsewhere in this conference. Here, we shall report mainly work carried out in Sussex on Vycor. Its aim has been to determine the disposition and motion of the adsorbed atoms and the nature of their magnetic interactions as a function of coverage, through observations of the nuclear magnetic susceptibility χ and the spin-lattice (T_1) and spin-spin (T_2) relaxation times. Both continuous wave and pulsed techniques have been used, at frequencies varying from 0.65 to 7.1 MHz. For coverages of a monolayer and below, the temperature range studied has been 0.3K to 2.5K. With higher coverages, measurements have been taken down to 45mK, and provide some interesting results on the nature of the interactions within these thicker films, and on the spin diffusion coefficient of Landau quasiparticles in limited geometries. These will not be discussed in detail here since they are outside the main scope of the conference.

During this work, we found[2] some important differences in the relaxation times, depending on whether or not the films had been annealed at 4K, and all of the results presented here refer to films in the fully annealed state.

NUCLEAR SUSCEPTIBILITY

The susceptibility is an essentially equilibrium observation which gives a measure of the magnitude and sign of the magnetic interaction between the nuclear spins. A collection of non-interacting spins in thermal equilibrium obeys Curie's Law, $\chi = C/T$, with the constant C having the free-spin value. This temperature dependence has been found to hold for a He^3 monolayer on Vycor porous glass[5], in the temperature range 0.3K - 2K. In Fig. 1, the Vycor results at T = 0.405K, are shown as total susceptibility signal versus amount of He^3 adsorbed. In these experiments although absolute values cannot be measured directly, the linearity of the graph in the submonolayer region indicates that the addition of each spin in this regime has no effect on the susceptibility of those already there, suggesting that the spin dependent interactions are much less than kT. In addition, there is good evidence[5] that at the higher coverages the behaviour is the same as that of bulk liquid. Absolute values of the susceptibility of the liquid have been determined by other workers by suitable normalisation procedures[6], and hence the data in Fig. 1 can be used to calculate the constant C, confirming that it has the free-spin value.

At lower temperatures, the susceptibility of coverages near one monolayer must depart from Curie's Law and might follow either a Curie-Weiss law, $\chi = C/(T + \Delta)$, appropriate to localised atoms, or a Fermi-like law, appropriate to a mobile two-dimensional Fermi fluid with a degeneracy temperature T_{2F}. It is perhaps more useful to compare one complete monolayer to the more localized system and the precision of the results in Fig. 1 allows one to place an upper limit on the exchange interaction Δ for this layer of $\sim$ 5mK. There is some evidence from measurements with filled pores that the second layer behaves like a two-dimensional Fermi fluid, and, the tendency towards parallel alignment of the spins indicates that the exchange interaction is considerably enhanced over that found in bulk liquid[7].

RELAXATION TIMES

By studying the way in which the magnetisation relaxes back to its equilibrium condition after it has been perturbed, we hope to be able to deduce the magnitude and the time variation of the local magnetic field since knowledge of these should allow us to predict the disposition and the physical state of the adsorbed atoms. There are two basic relaxation times about which we would like information; one of these is T_1, the spin lattice relaxation time, which is the characteristic time taken for the

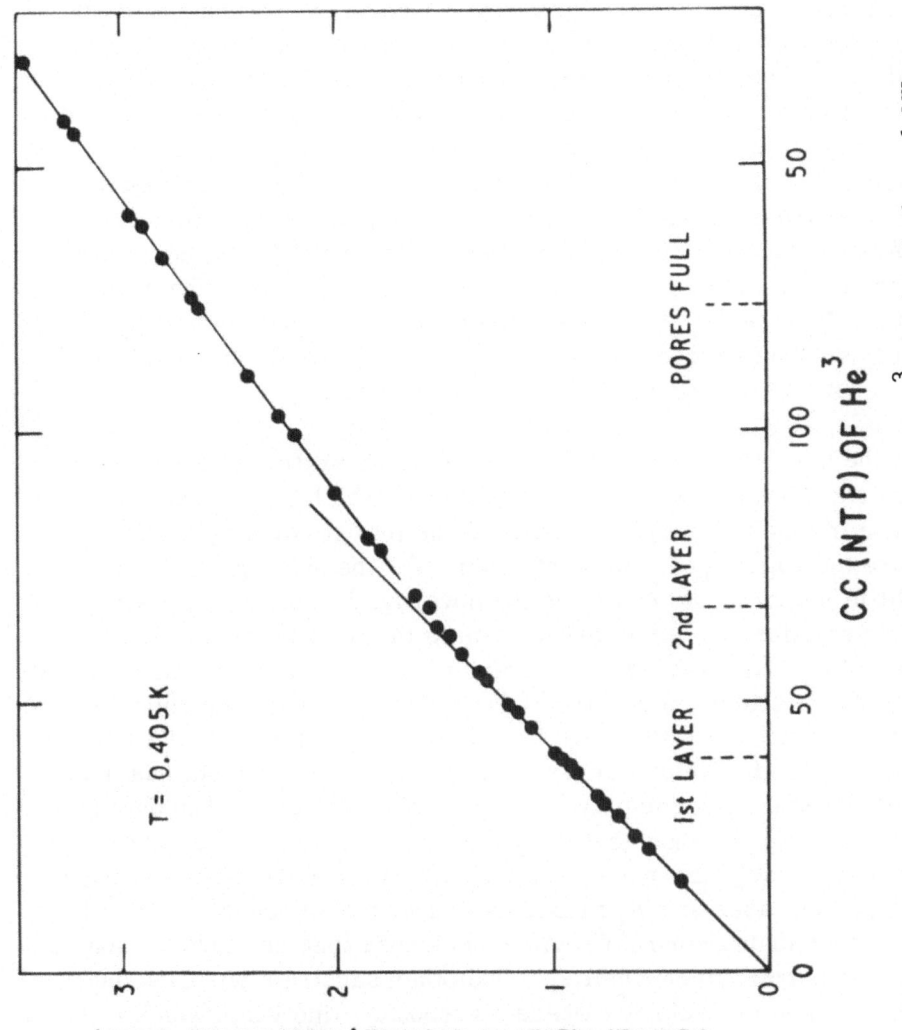

Fig. 1. Susceptibility versus quantity of ^{3}He adsorbed at 6 MHz.

magnetisation of the spin system (i.e. population difference between different spin states) to recover after it has been disturbed and is determined by the strength of the forces between the spins and the lattice in which they reside; the other time is T_2, the spin-spin relaxation time, which is the characteristic time taken for the precessing spins in a magnetic field to dephase relative to each other and is determined by the range of local magnetic fields present. A considerable change occurs in these relaxation times as the amount of adsorbed helium is increased and this effect can be seen in Fig. 2 where T_1 and T_2 are plotted as a function of θ, the fractional monolayer coverage. The T_1 data in this figure were all measured at a frequency of 6.24 MHz by saturating a continuous wave signal and observing its recovery. The T_2 data derive from two sources with some of the points being taken from the 6.24 MHz data by a combination of measuring line widths and saturation factors while the points taken at 2.14 MHz were found using spin echo measurements involving 90° - 180° pulse sequences. It is notable that for the smaller coverages these two sets of T_2 agree well, giving confidence in the results of the two methods of measurement.

In the standard BPP theory of magnetic relaxation of Bloembergen, Purcell and Pound[8] the larger the factor by which T_1 exceeds T_2, then the more immobile the spins. Thus at the lowest coverages where we measure the ratio T_1/T_2 to be of order 10^6, the spins probably have very little motion. However, as the coverage increases this ratio becomes smaller and our interpretation is that this is due to the local magnetic field experienced by the spins undergoing a faster variation with time. There are not many other adsorbates on which such relaxation time data for adsorbed He^3 have been obtained but some results have been obtained on Zeolite which like Vycor glass is a dielectric but has much smaller pores ($\sim 10 \text{Å}$ compared with $\sim 60 \text{Å}$ for Vycor). The data of Monod and Cowen[1] show that the relaxation times on this substrate behave in precisely the opposite way to the data presented in this paper in that T_1 increases and T_2 reduces as coverage increases. Thus it would appear that adsorbed helium experiences less and less motion as a monolayer is built up on Zeolite. One other substrate on which data have been taken is Grafoil[3] where a resonance line width which reduces with increasing coverage indicates similar motional narrowing to the kind presented in this paper.

The question that arises is how to quantify this motion and to calculate its correlation time. Our estimate of this time comes from studying the frequency dependence of T_2 shown in Fig. 3. T_2 should vary with frequency only in the region where the correlation time τ for the local

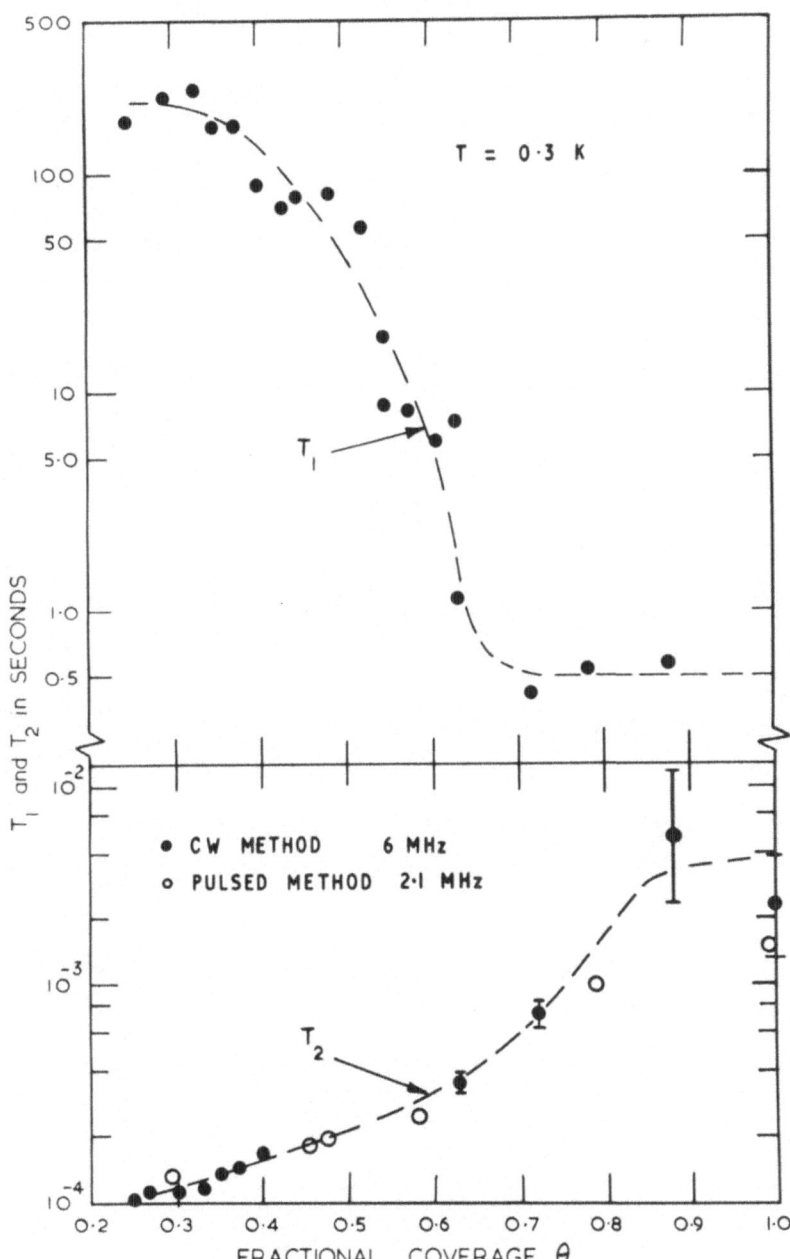

Fig. 2. Spin–lattice and spin–spin relaxation times versus fractional coverage

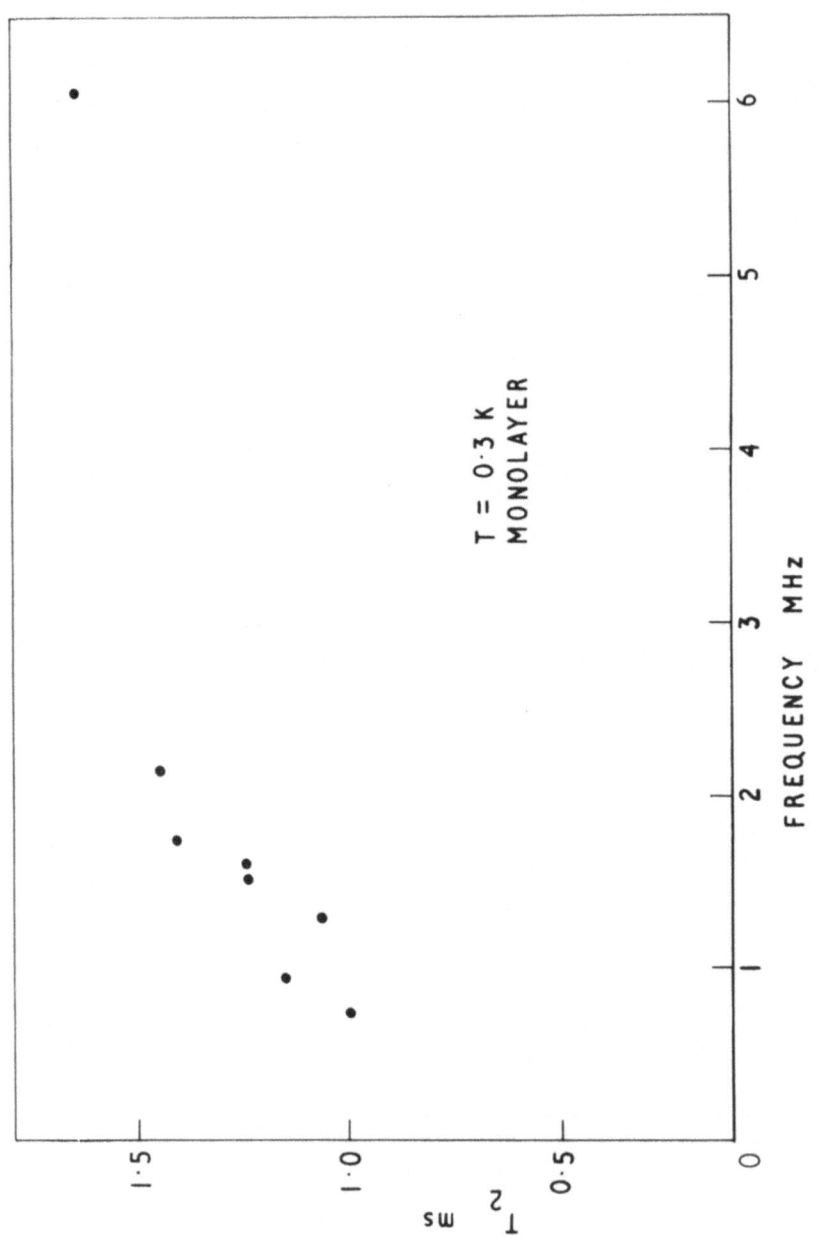

Fig. 3. Spin-spin relaxation time versus frequency.

field fluctuations is of the order of ω^{-1}, the reciprocal of the Larmor frequency.   The variation observed in Fig. 3 suggests therefore that τ has a value of $\sim 10^{-7}$ secs.

The nature of this motion could perhaps be determined by observing how the relaxation times change with other variables.   In Fig. 4 the relaxation times taken by the continuous wave method are plotted as a function of temperature for a complete monolayer and two partial coverages. The values are very nearly constant and a tentative interpretation of this is that most of the motion present in the spin system is not thermally activated but is caused perhaps by quantum tunnelling.   This tunnelling might either be through substrate potential barriers or else with other helium atoms due to an exchange interaction.   Another possible interpretation is that if the measurements were indeed made in the region where $\omega \tau \sim 1$, then T_1 ought to be near a minimum in its value measured as a function of τ and therefore alterations in the temperature might be expected to effect it to a reduced extent.

So far we have presented the data in which T_1 was derived from continuous wave measurements for which the recovery of the resonance line was always observed to be exponential.   However spin-echo measurements allowed a more precise measurement of the recovery of the magnetisation particularly at short times after disturbing the system and typical graphs of the recovery of the signal are shown logarithmically in Fig. 5 for several different frequencies at a monolayer coverage. These data were obtained by observing either the free induction decay after a 180° - 90° pulse sequence or the echo after a 180° - 90° - 180° sequence.   It is notable that these plots show the recovery is not given by a simple exponential and furthermore that qualitatively similar recoveries were observed for the fractional coverages.   The slope of the initial recovery is approximately one quarter of the last part and presumably is being caused by some mechanism more complicated than normally met in NMR studies where exponential recoveries for T_1 are usually found.   One possible reason for this behaviour is that the spins exist in a range of different adsorbent states possessing different interactions with the lattice and hence a variety of spin lattice times. Another reason may be that while the spin system might be fairly homogeneous there may be more than one process by which the spins can give energy to the spin lattice.   This happens in the case of solid ^{3}He [9] where alongside the direct process of Zeeman energy being injected into the lattice there is another process in which Zeeman energy is converted into energy associated with ^{3}He spins exchanging places and subsequently quanta of energy appropriate to this exchange flow into the lattice.   The

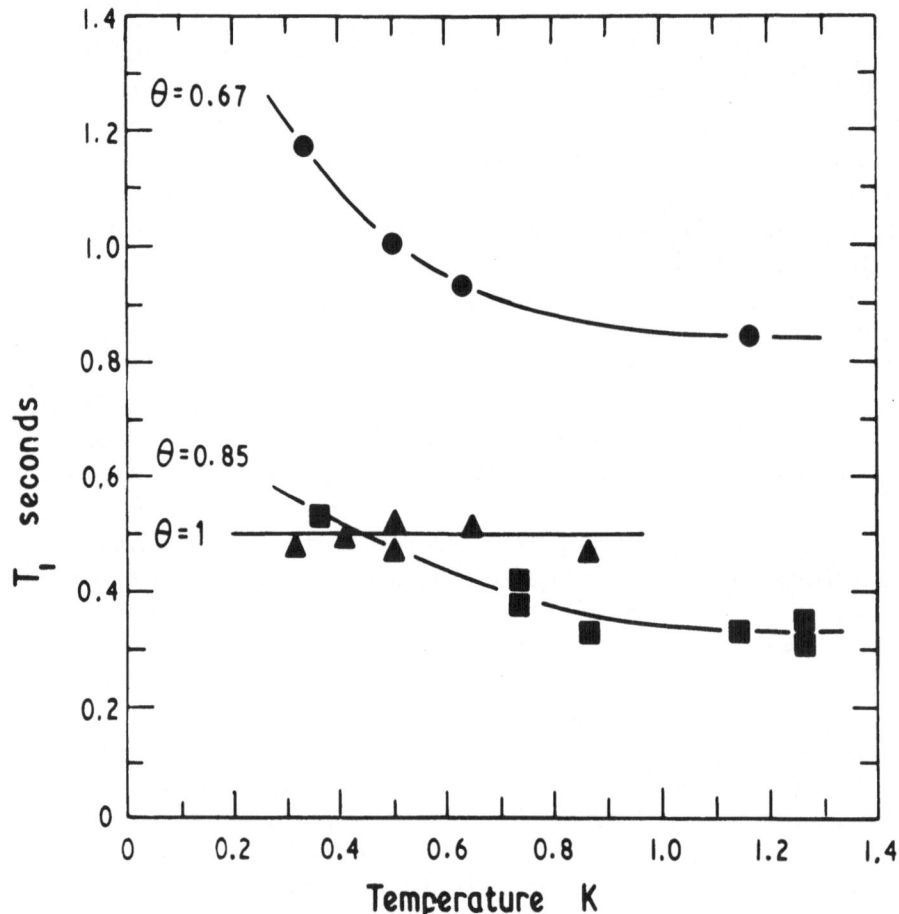

Fig. 4. Spin-lattice relaxation time versus temperature
at 6 MHz.

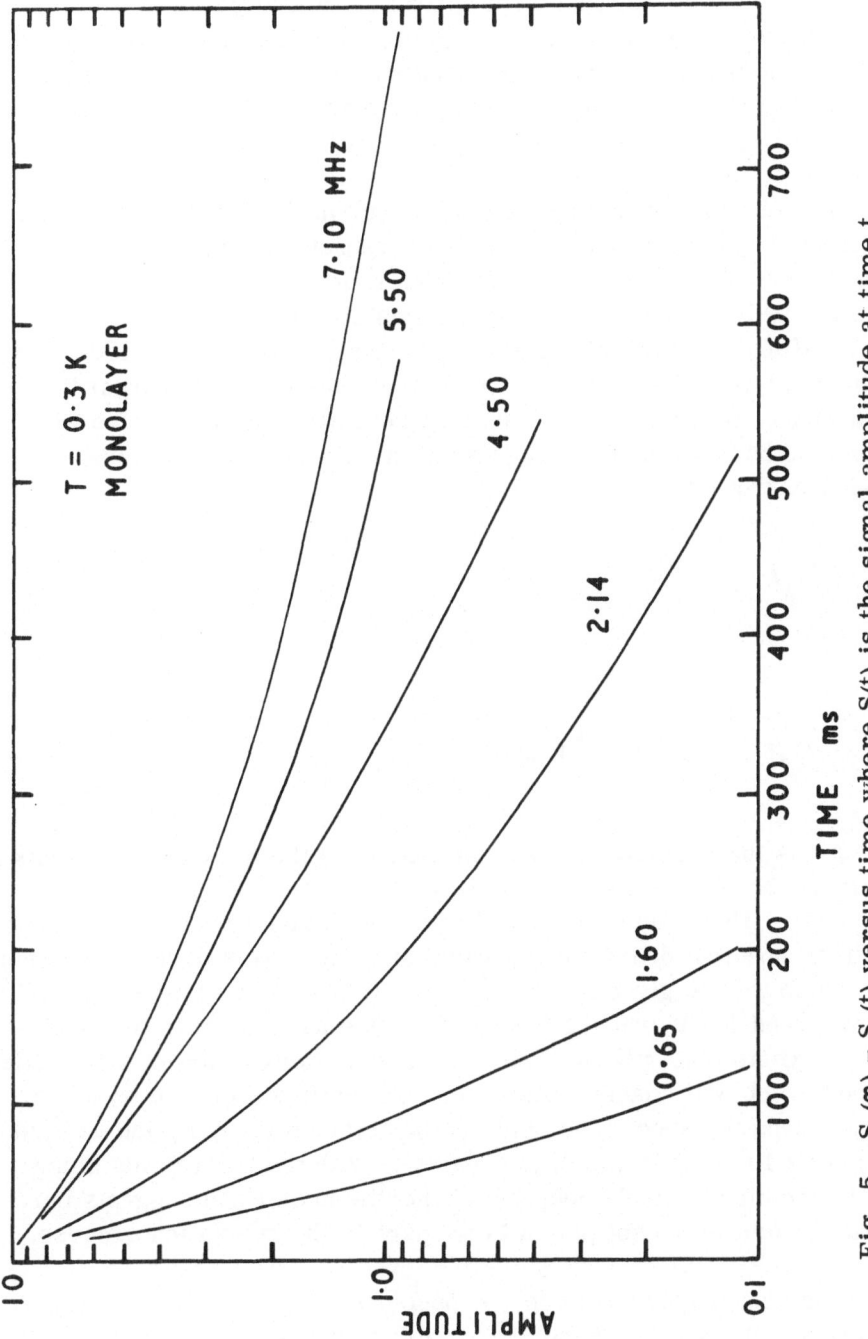

Fig. 5. S (∞) – S (t) versus time where S(t) is the signal amplitude at time t.

second explanation would imply that the data should be analysed as a sum
of two exponentials rather than as a spread of relaxation times and
within the accuracy of our data either interpretation is possible.   In the
case of T_2, the observed results were always exponential but at the lower
coverages, the initial part of the relaxation was lost due to receiver
recovery.

Another obvious property that can be seen in the family of curves of
Fig. 5 is that T_1 varies significantly with frequency.   Plotted in Fig. 6,
as a function of frequency is the relaxation time at the start of the recovery
and this is labelled as T_1'.   The approximately linear variation of T_1'
with ω is difficult to account for although a similar linear dependence has
been observed for the spin-lattice relaxation time by other workers[1]
and is probably characteristic of surface relaxation.   In any case we
can form another estimate of τ by noting that provided the resonance is
motionally narrowed,

$$\frac{1}{T_2} \sim M_2 \tau$$

and

$$\frac{1}{T_1} \sim M_2 \tau \left[\frac{1}{5} j \ (\omega) + \frac{4}{5} j \ (2\omega) \right]$$

where $j \ (\omega)$ is the normalized spectral density of the motion of the spins in
the lattice.   Since it is normalized we expect that $j \ (0) = 1$, $j \ (\infty) = 0$
and $j \ (\frac{1}{\tau}) \sim \frac{1}{2}$.   Hence an estimate of τ can be obtained from the frequency
at which the spectral density has a value of $\frac{1}{2}$ and at this point the value of
T_2/T_1 , which will be given by $\frac{1}{5} j \ (\omega) + \frac{4}{5} j \ (2\omega)$, will be of order 0.1.
The observed T_2/T_1 ratio rises to this value as $\omega/2\pi$ falls to ~ 1 MHz
and this suggests that $\tau \gtrsim 10^{-7}$ secs.   Further work at lower frequencies
is planned which will enable a more accurate estimate of τ to be made.
If $\tau \sim 10^{-7}$ secs the second moment can be obtained from T_2 and we find
that $M_2 \sim 2 \times 10^8$ secs^{-2} which is consistent with the local fields being
caused by the dipole-dipole coupling of the ^{3}He nuclei and suggests paramagnetic impurities do not play a large part in the relaxation process.

At lower coverages the ratio $\frac{T_2}{T_1}$ becomes progressively smaller
indicating that the motion of the spins is reduced.   In the limit of
the lowest coverages which we have measured ($\sim \frac{1}{10}$ monolayer) it is
probable that the line width has achieved the value appropriate for a rigid
lattice when $T_2 \sim M_2^{-\frac{1}{2}}$, the inverse of the square root of the second

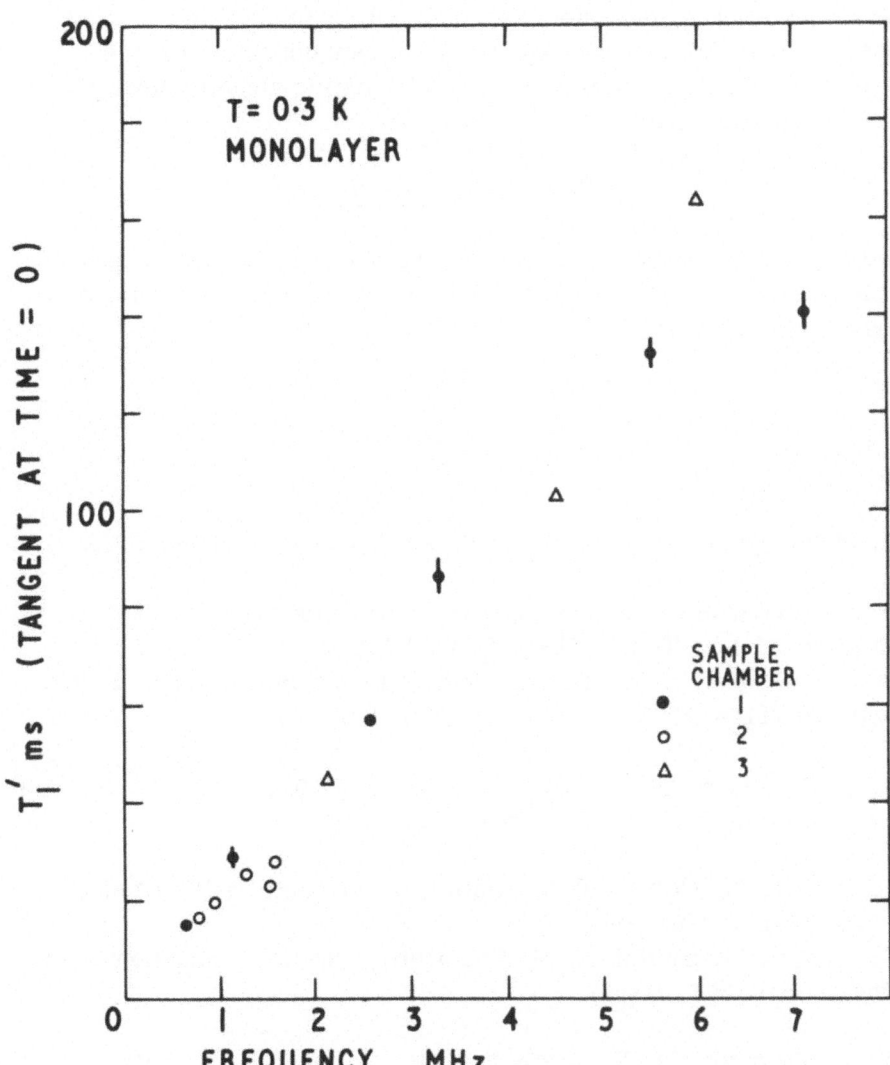

Fig. 6. Initial spin–lattice relaxation time versus
frequency.

moment.   The value of T_2 was recorded here to be ~ 70 microsec suggesting therefore a value for M_2 of $\sim 2 \times 10^8$ sec^{-2}.   Within the accuracy of our measurements and interpretation it is coincidental that this is the same value as predicted above for the monolayer but at least it seems likely that M_2 does not change rapidly with coverage.   Thus it is possible that the formation of the monolayer consists of rather tightly packed islands growing in size and becoming steadily more mobile in character as the island size increases.

ACKNOWLEDGEMENTS

This work has been supported in part by the Science Research Council (Grant No. B/RG/1795), and by the U.S. Army through its European Research Office (Contract No. DAJA 37-73-C-2416).

(1) P. Monod and J. A. Cowen, unpublished private communication.

(2) For a review of these experiments, with references, see D. F. Brewer, J. Low Temp. Phys. 3, 205 (1970)
D. J. Creswell, D. F. Brewer and A. L. Thomson, Phys. Rev. Lett. 29, 1144 (1972).

(3) R. J. Rollefson, Phys. Rev. Lett. 29, 410 (1972).   D. P. Grimmer, Proc. 13th Int. Low Temp. Phys. Conf. (1972).

(4) G. Careri, M. Guira and M. Santini, Phys. Lett. 4, 61 (1963); Phys. Lett. 5, 102 (1963).
G. Careri, M. Santini and G. Signorelli, Proc. 9th Int. Low Temp. Phys. Conf. 364, (1965).

(5) A. L. Thomson, D. F. Brewer and P. J. Reed, Proc. 10th Int. Conf. Low Temp. Phys., 338 (1968).
D. F. Brewer, D. J. Creswell and A. L. Thomson, Proc. 12th Int. Conf. Low Temp. Phys., 157 (1970).

(6) J. R. Thompson, H. Ramm, J. F. Jarvis and H. Meyer, J. Low Temp. Phys. 2, 521 (1970).
H. Ramm, P. Pedroni, J. R. Thompson and H. Meyer, J. Low Temp. Phys. 2, 539 (1970).

(7) D. F. Brewer and J. S. Rolt, Phys. Rev. Lett. <u>29</u>, 1485 (1972)

(8) N. Bloemberger, E. M. Purcell and R. V. Pound, Phys. Rev. <u>73</u>, 679 (1948).

(9) M. G. Richards, N. M. R. in Helium Three (Advances in Magnetic Resonance, Vol. 5, Ed. J. S. Waugh, Academic Press, New York and London, page 305, 1971).

c.w. NMR OF HE^3 ON GRAPHITE

R. J. Rollefson

Department of Physics, University of Washington

Seattle, Washington 98195

I INTRODUCTION

Recently there has been considerable interest in the study of helium films adsorbed on graphite. For helium coverages in the first monolayer early specific heat (1) and NMR measurements (2) gave a consistent and rather detailed picture of adatom behavior. Recent results (3), however, have indicated that the situation with respect to the NMR may be somewhat more complex than originally envisioned. In hopes of gaining greater insight into this problem the early NMR results are here reviewed with particular attention given to the experimental details.

II EXPERIMENTAL TECHNIQUE

The experiments were performed using graphitized carbon black (4) as a substrate. This type of graphite is in the form of small polyhedra $\sim 0.2\mu m$ in diameter with a layered, onion-like structure exposing only basal surfaces. In the graphitized carbon black as received the graphite particles were stuck together into small grains ranging in size up to about two millimeters in diameter. To improve the filling factor of the sample cell these grains were broken up by pressing them through a fine brass screen. The resulting fine powder was poured into the sample cell and allowed to settle but not forcefully packed.

The quartz sample cell with a fill line running to a shutoff valve at the cryostat top plate was mounted in the cryostat in such a way that the entire graphitized carbon black sample could

be baked at 1000°C. During the bake-out of about 50 hours the cell
was flushed several times with He⁴ gas purified by passing through
a liquid nitrogen trap containing Union Carbide 13-X molecular sieve.
It was otherwise kept under vacuum by continual pumping using the
same cold trap for protection. At the end of the bake-out the cell
was isolated by closing the shut-off valve while the sample was
still hot. A close fitting NMR coil was greased to the outside of
the quartz tube and the cryostat mounted in the dewars. In this
manner the substrate was kept under vacuum at all times between the
bake-out and the introduction of the He³ adsorbate.

The NMR absorption was measured with a standard Rollin's spec-
trometer operating at 20.5 MHz using phase sensitive detection.
Line widths were measured as the difference in field between the
maximum and minimum points of the absorption derivative signal.
To improve the signal to noise ten curves were taken at each temperature and coverage and the widths averaged to give the data point.

The following procedure was used in taking data: liquid He⁴
was transferred into the dewar and the required amount of He³ gas
admitted to the sample chamber through the 13-X molecular sieve
cold trap. A resistence thermometer thermaly anchored to the NMR
coil indicated that the coil, and thus the quartz sample tube,
reached 4.2K in about 15 minutes. The cryostat was then left at
4.2K overnight before beginning data taking. A run at a particular
coverage lasted from three to four days with the data being taken
at progressively lower temperatures. After the data at a given
temperature was complete the bath was pumped to the next lower
temperature and the sample allowed to equilibrate for at least one
hour. Except for liquid helium transfers (lasting less than 15
minutes) the bath was never allowed to warm up during a run.

In regions where the line width was changing rapidly with
temperature the measured width served as a reasonably good thermometer, changing significantly from one temperature to the next. As
a check on whether the sample was reaching equilibrium in the one
hour time allowed some points were taken after a considerably longer time ranging from 16 to 26 hours. No evidence was ever found of
a further decrease in sample temperature (i.e. increase in line
width) for these longer times. In particular before taking the
lowest temperature point of each run the sample temperature was
held for at least 16 hours.

III DATA

The line width as a function of temperature and coverage is
shown in Fig. 1. At high temperatures the line attains a width of
about 0.32 Gauss, independent of coverage. The applied magnetic

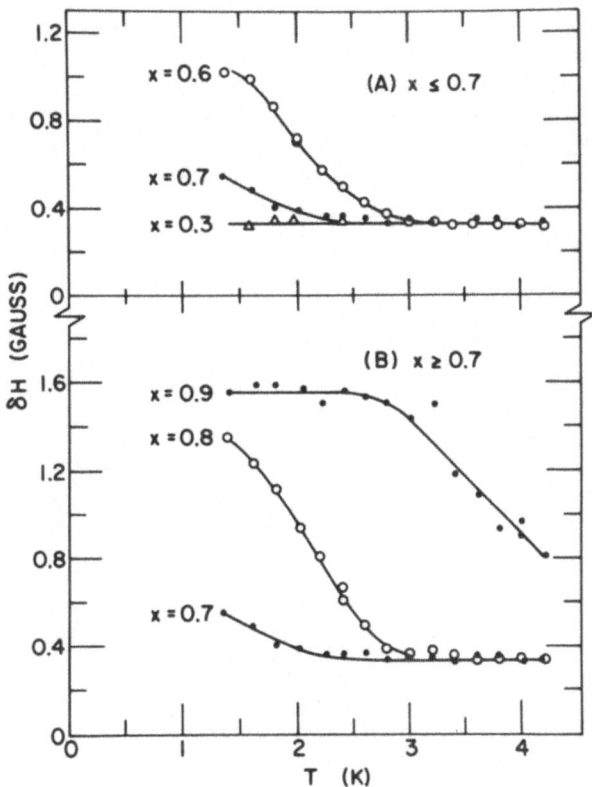

Fig. 1 Experimental line width as a function of temperature.
x = fractional He³ monolayer coverage.

field had a homogeneity of about 0.1 Gauss (5). Since the measured
width is independent of coverage it is unlikely to be due to helium-
helium interactions. For the purposes of this paper we will simply
consider it as an effective field inhomogeneity.

At lower temperatures the line width changed in a manner which
was strongly dependent on temperature and coverage. For a fractional coverage, x, of 0.3 no change in width was observed, and for
x = 0.7 only a small increase occurred. On the other hand for
x = 0.6, 0.8, and 0.9 rather abrupt width increases resulted with
the temperature of the increase dependent on coverage. For x = 0.9

a temperature independent width of about 1.6 Gauss was attained.
Both 0.8 and 0.6 monolayer coverages were still increasing in width
at the lowest temperature but appeared to be bending toward temperature independent widths.

For a Curie law susceptability, which at these temperatures
should be a good approximation to He^3 behavior, the static susceptibility (proportional to the area under the absorption curve) varies
as the inverse temperature. If the line shape $g(\omega)$ has a constant
functional form then the areas under $g(\omega)$ will be proportional to
the maximum of the derivative times the square of the line width.
These two parameters could be obtained directly from the output
data. In Fig. 2 the quantity $g'_{MAX}\Delta^2/x$ is plotted versus T^{-1} where

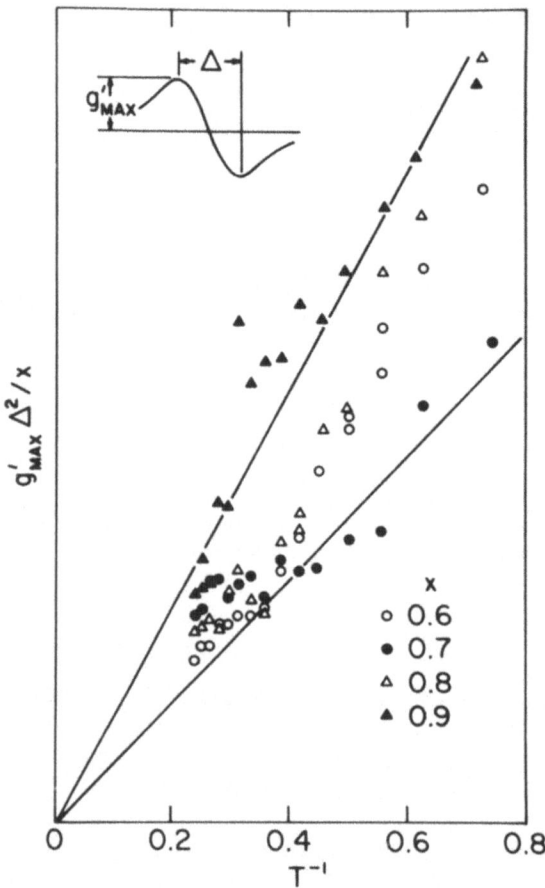

Fig. 2   Curie law test of absorption curve area versus inverse
         temperature.

Δ is the width between the maximum and minimum of the derivative
and g_{MAX} is the maximum value of the derivative. The factor x
allows for the change in the number of He^3 atoms with coverage.
Although the slope changes for different coverages as indicated by
the lines for x = 0.7 and 0.9, for a given coverage proportionality
to T^{-1} is observed. This is taken as further evidence that the
sample was in thermal equilibrium during the measurements. As noted
above the quantity plotted is only proportional to the area if the
line shape remains the same. If, on the other hand, a line of
given area were to change from a Lorentzian to a Gaussian shape the
quantity plotted in Fig. 3 would increase by a factor of 3.5.

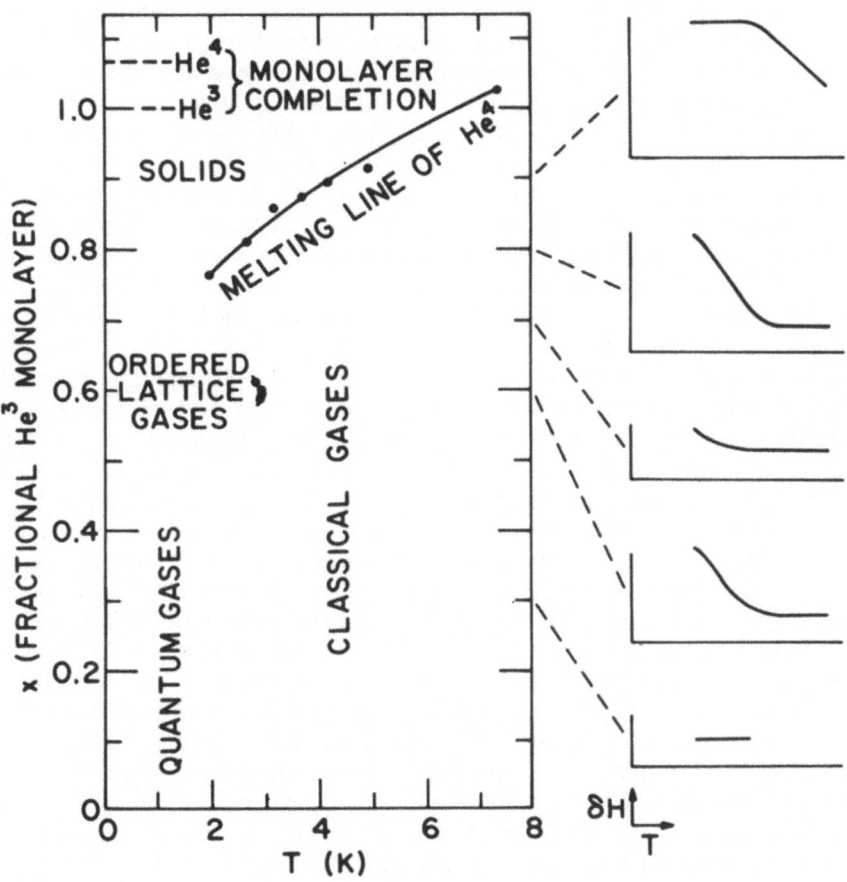

Fig. 3 Phase diagram for monolayer helium films. Data points de-
        note peaks in the specific heat. The diagrams to the right
        are schematic representations of the data in Figure 1. Al-
        though the melting line shown is for He^4, the heat capacity
        of He^3 with x ≃ 1 was measured, showing similar behavior.

This is because a Lorentzian line has more area in the wings. Thus a possible explanation for the difference in slope is that the line for x = 0.7 is more Lorentzian in character while that for x = 0.9 is more Gaussian.

A further note should be added at this point concerning the question of thermal equilibrium. It is well known that in powder samples at low temperatures thermal equilibrium is difficult to obtain. Considerable evidence has been presented above indicating that in the experiments reported here thermal equilibrium was indeed achieved: 1) There is a strong temperature dependence of the line width, indicating that the sample is cooling, and this dependence changes in a regular way with coverage; 2) the response time of the sample temperature, as measured by the line width in a region of rapid change with temperature, was no more than one hour; 3) the sample susceptability, as measured by the area under the absorption curve, showed the expected T^{-1} increase. However, such behavior has not always been the case. After the measurements reported here were completed the system was evacuated, closed off, and left. Recently the apparatus was started up again with the intent of extending the measurements to a greater number of coverages. During the months when the apparatus had been closed off gas from the walls of the capillary, etc., had accumulated in the sample chamber. This was pumped to a good vacuum before cooling the sample. Data was taken at x = 0.65, 0.75, and 0.85. The line width observed was equal to about 0.3 Gauss and independent of the temperature as measured by both the pressure and the resistance thermometer in contact with the NMR coil. The area under the absorption surve was proportional to coverage (i.e. to the number of helium atoms) but was also independent of temperature. Thus the evidence is that the sample in this second set of measurements was not cooling as the temperature was lowered. The sample was re-baked in a manner identical to the original bake-out and measurements were again made at x = 0.85. Again no temperature dependence was observed in the line width or curve area. The cryostat was then disassembled and the sample inspected. No change was apparent. At present the reason for the discrepancy between these two sets of runs is not known and is the subject of further investigation. However, the evidence is strongly in favor of the validity of the first set of runs, the results of which are shown in Fig. 1 and Fig. 2. While the apparatus was dormant evidently some unknown problem occurred which has resulted in the difficulty with the more recent set of runs.

IV DISCUSSION

The decrease in line width with increasing temperature shown in Fig. 1 is interpreted as reflecting an increase in mobility,

the well known motional narrowing. The line width at low temperatures is caused by a variation in the local field seen by the He³
atoms. When the diffusive excursions of the atoms are sufficiently
large to cause a significant change in the local field during the
time of a Larmour precession in this field (in the present case
$\sim 10^{-5}$ sec.) the local field is effectively averaged and the line
narrows. Thus the data of Fig. 1 show two coverage regions (x = 0.6
and x = 0.8) in which the low temperature mobility is reduced with
a region between (x = 0.7) in which there is relatively higher
mobility. These observations are in agreement with heat capacity
measurements and have been interpreted as showing 1) an ordering
of the adsorbate atoms on the substrate potential at coverages
around x = 0.6 and 2) a transition to a 2D "solid" phase for coverages approaching one monolayer. The agreement between the NMR results and the heat capacity results is shown in the phase diagram
of Fig. 3.

It might be noted that for motionally narrowed lines such as
x = 0.7 one expects a Lorentzian like line shape while for broadened rigid lattice lines such as x = 0.9 the shape is more Gaussian
in character. These are the same shapes that were proposed to explain the different slopes in Fig. 2.

The question remains as to the origin of the local field pro-
ducing the broadened lines a low temperature. It is well known
that in solid bulk helium the large zero-point motion and resulting
overlap of the atomic wave functions allows the He³ atoms to tunnel
around each other resulting in an "exchange narrowing" of the NMR
line. In the present experiments the helium atoms are constrained
to move on a plane which due to the hard cores greatly restricts
their ability to interchange positions. It is possible that this
"steric hindrance" results in the suppression of exchange narrowing
for coverages near one monolayer while the registry with the substrate potential reduces exchange for coverages near 0.6 monolayers.
Thus the full rigid lattice line width caused by the dipole interactions may be observed. If this is so then the maximum low temperature width should depend on coverage, ie. on interatomic spacing.
By the use of second moment formalism it is possible to calculate
the line width due to dipole-dipole interactions. This was done
for the helium films with the assumption that the helium atoms are
in a uniform triangular array, allowing the calculation of the
interatomic spacing from the measured substrate surface area, and
that the line shape is Gaussian, allowing the width to be related
to the second moment. The effect of the 0.32 Gauss "field inhomogeneity" was accounted for by adding the square of this width to
the square of the calculated dipolar width. The results are:

Fractional coverage	Maximum observed width	Calculated width
0.6	1.0 Gauss	1.0 Gauss
0.8	1.35 "	1.5 "
0.9	1.6 "	1.75 "

The maximum observed widths are consistant with those which would
be expected for a rigid lattice dipole interaction. However, it
must be noted that while the line widths for x = 0.6 and 0.8 appear
to be leveling off at low temperatures, it is possible that the
maximum line width could be significantly larger than that observed.
In this case the question of the origin of the local fields responsible for the low temperature line width would have to be
re-examined.

REFERENCES

1. M. Bretz and J. G. Dash, Phys. Rev. Lett. 27, 647 (1971).
    M. Bretz, G. B. Huff, and J. G. Dash, Phys. Rev. Lett. 28, 729
    (1972).
2. R. J. Rollefson, Phys. Rev. Lett. 29, 410 (1972).
3. D. C. Hickernell and D. P. Grimmer, this conference.
4. Grade Sterling FT, Cabot corporation, Boston, Mass.
5. Misallignment of the magnet with respect to the sample would
    of course result in a greater inhomogeneity. Although the
    weak signal made allignment of the magnet somewhat difficult,
    it seems unlikely that this could be the sole cause of the ob-
    served line width. Furthermore, the magnet has been moved and
    repositioned several times with the minimum line width always
    being the same.

PULSED NMR IN He^3 ADSORBED ON GRAPHITE BELOW 4.2K[*]

D. P. Grimmer and K. Luszczynski

Department of Physics, Washington University

St. Louis, Missouri 63130

Pulsed NMR experiments have been performed on He^3 adsorbed on pyrolytic graphite. Most of the work has been done with Grafoil[1] substrates, but Sterling FT[2] has also been used in a few experiments. Submonolayer, monolayer, and multilayer coverages have been examined. The temperaure range of the experiments was from 0.35K to 4.2K. The results described here have not been reported previously.[3]

The main purpose of our experiments is to determine the nature of the adsorbed He^3 system using the nuclear spin system as a probe. NMR in adsorbed He^3 provides information about the spin system and and the magnetic environment of the nuclei, and hence about the adsorbed system as a whole.

The construction of the Grafoil substrates is similar to what has been reported previously,[3] but with some modification (Fig.1): a strip of 0.010 inch thick Grafoil sheet, together with a strip of 0.001 inch Teflon sheet as electrical insulator, was rolled into a cylindrical spiral around a sapphire rod 0.1 inch in diameter, and slipped into a pyrex collar with an internal diameter of 0.53 inch. The substrate was bonded with GE 7031 varnish to the sapphire rod, and the sapphire rod and substrate bonded to a 0.50 inch x 0.625 inch diameter sapphire disk. The radio frequency magnetic field (H$_1$) was applied along the axis of the rod. This configuration was used to reduce eddy currents in the Grafoil substrate, caused by the radio frequency (rf) pulses. The sapphire rod and plate system also provided good thermal contact to a He^3 refrigerator. The entire substrate assembly was enclosed in a pyrex tube attached to a copper plug by means of a Housekeeper seal. A tapered collet on

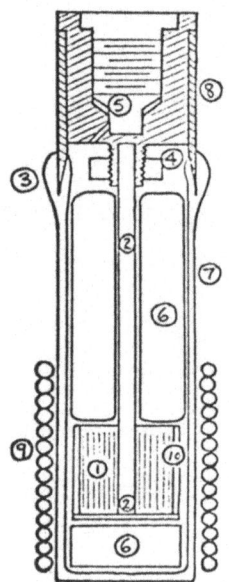

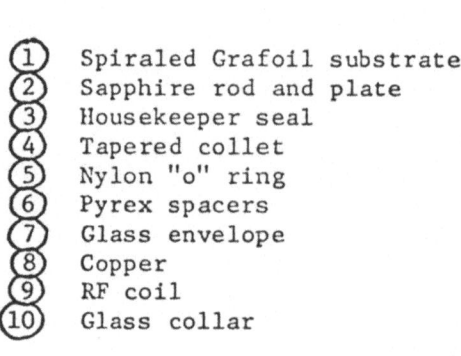

① Spiraled Grafoil substrate
② Sapphire rod and plate
③ Housekeeper seal
④ Tapered collet
⑤ Nylon "o" ring
⑥ Pyrex spacers
⑦ Glass envelope
⑧ Copper
⑨ RF coil
⑩ Glass collar

Figure 1. Sample chamber and Grafoil substrate.

the plug held the sapphire rod and the plug was screwed to the copper He³ refrigerator on the bottom of the low temperature probe. A special contraction seal using a nylon "o" ring permitted the superfluid-tight joining of the sample chamber to the probe. This removable sample chamber represents the major change from the sealed-bomb sample chamber used in earlier experiments.[3]

The spiraled substrate using GE7031 varnish as a cement has the advantage of maximum packing density in a cylindrical sample chamber. However, it also has several disadvantages: (1) the substrate cannot be baked above 150°C, and (2) electrical and magnetic anisotropies present in Grafoil are averaged, making interpretation of the data more difficult.

Adsorption isotherm measurements were performed on different substrates made with Grafoil (Fig. 2). Most adsorption isotherms were made using argon as the adsorbate at 77K, but for one substrate an isotherm for He³ adsorbed at 4.2K was also measured. The monolayer capacities, determined in the usual manner from the first inflection point on the stepwise isotherms, indicates a specific area of 20 m²/gram and an area per He³ atom of 9.95A². The area per atom for argon is taken to be 12.8A².[4] In both isotherm measurements and adsorbed He³ NMR experiments, the sample chamber was heated to ~130°C and evacuated with a diffusion pump for ~70 hours.

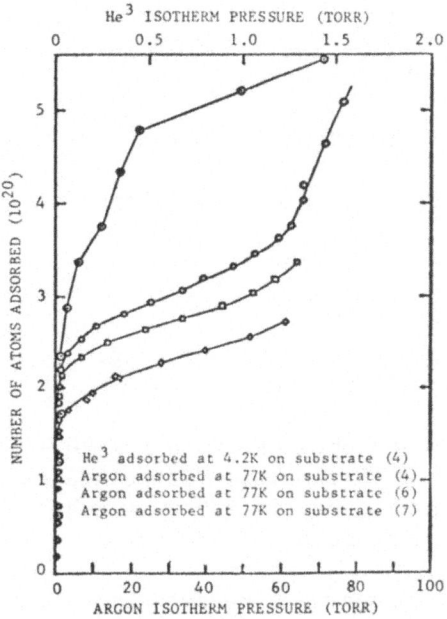

Figure 2. Adsorption isotherms for He³ (at 4.2K) and argon (at 77K) on various Grafoil substrates differing in the amount of Grafoil and construction technique.

Pulsed NMR techniques were employed. Most data were taken at 20 MHz but some experiments were done using 10 MHz resonant frequency. The finite electrical conductivity of the Grafoil and the resulting eddy current heating presented the major experimental difficulty. Care was taken to achieve thermal equilibrium after each pulse sequence, so that the results represent the equilibrium situation. The following phenomena were investigated in our experiments: the free induction decay (FID); spin-spin (T_2) times; spin-lattice (T_1) times; the nuclear magnetic susceptibility (χ); and the effect of spin-diffusion in a linear magnetic field gradient. Coverages explored range from 0.1 monolayer through multilayer (up to ~21% saturated pores). In addition to pure He³ coverages, a He³ monolayer mixed with varying amounts of He⁴ on Grafoil was examined. NMR measurements on He³ adsorbed on Sterling FT were also made for comparison with the Grafoil data.

Except for multilayer coverages the FID is exponential. The decay times observed are considerably shorter (~100μs) than the magnetic field inhomogeneity decay time of about 4ms obtained with a comparable bulk liquid glycerine sample. The observed FID in adsorbed He³ is not the real T_2 (~1ms), which is measured from the echo envelope produced by multiple 90°-180° pulse sequences. Hence, the dephasing of spins controlling the FID is not random, since the spins can be rephased by a 180° pulse to produce an echo. The effect is one like a macroscopic magnetic field inhomogeneity

of the order of 0.3 gauss/cm. The shape of the observed FID corresponds to a Lorentzian line. This broadened Lorentzian line is
believed to be caused by the macroscopic magnetization arising
from the diamagnetism of the Grafoil.

The T_2 relaxation time behavior for He^3 on Grafoil can be summarized as follows:
(1) T_2-decay is generally exponential except for the saturated
coverage system.
(2) T_2 increases with increasing temperature.
(3) T_2 increases with increasing multilayer coverage x very sharply at 4.2K, less so at 1.16K. T_2 for the submonolayer coverages
is essentially independent of coverage (Fig. 3).
(4) T_2 was longer for 10 MHz than for 20 MHz, by about a factor of
1.8.
(5) The width of the echo obtained with the $90°$-τ-$180°$ pulse sequence increased with τ for multilayer coverages.
(6) The one monolayer (ML) He^3 in He^4 system has a T_2 about the
same as the corresponding pure He^3 system.

The T_1 relaxation times were measured mostly at 4.2K and 1.16K.
The data can be summarized as follows:
(1) For submonolayer coverages T_1 lengthens as the temperature is

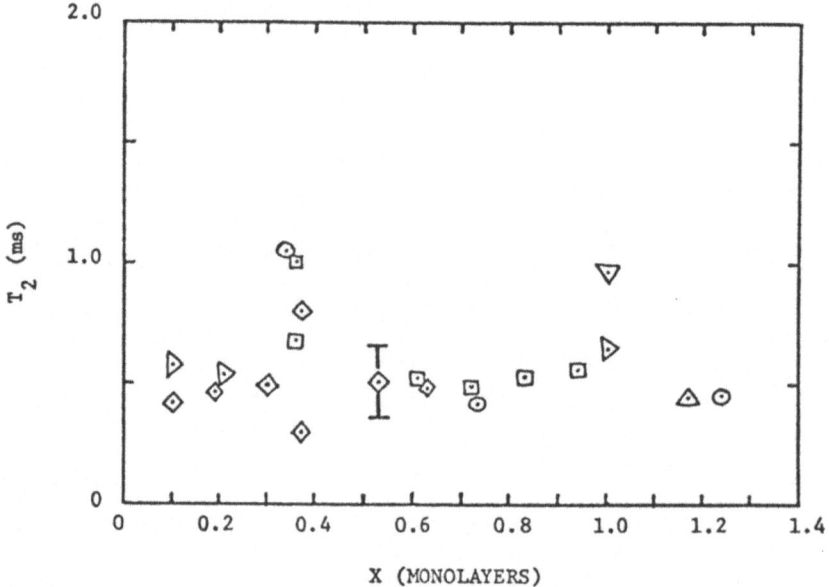

Figure 3. Spin-spin relaxation time (T_2) vs. submonolayer coverage
(x) for He^3 adsorbed on Grafoil at T=1.16K.

lowered. However, the T_1 for multilayer coverage is rather insensitive to temperature.
(2) There is an increase in T_1 with multilayer coverage.
(3) T_1 was shorter for 10 MHz than for 20 MHz at the same coverage.
(4) The one ML He³ in He⁴ system T_1 is shorter than corresponding
coverage pure He³ system at 1.16K and about the same at 4.2K.

The observed nuclear magnetic susceptibility (χ) follows
Curie's law down to about 1K, below which temperature the magnetization apparently falls below that predicted by Curie's law.
This was observed for submonolayer, monolayer, and thick layer
(21% saturated-pore) coverages (Fig. 4). The nuclear magnetization of He³ at a given temperature is measured by fitting a FID
curve to an exponential and extrapolating this curve back to the
beginning of the decay, that is the end of the rf pulse. It
should be noted that in bulk He³ the effective Fermi degeneracy
temperature T_0 varies from 0.45K in liquid He³ at saturated vapor
pressure to 0.26K as the solid density is approached. On solidification T_0 drops down to several millidegrees. The separation between the He³ atoms in the adsorbed monolayer corresponds to that
obtained in bulk solid He³ with a molar volume of about 20 cm³ and
T_0 in the range of 6 mk. For the saturated-pore coverage the fall-
off in susceptibility from Curie's law values at lower temperatures

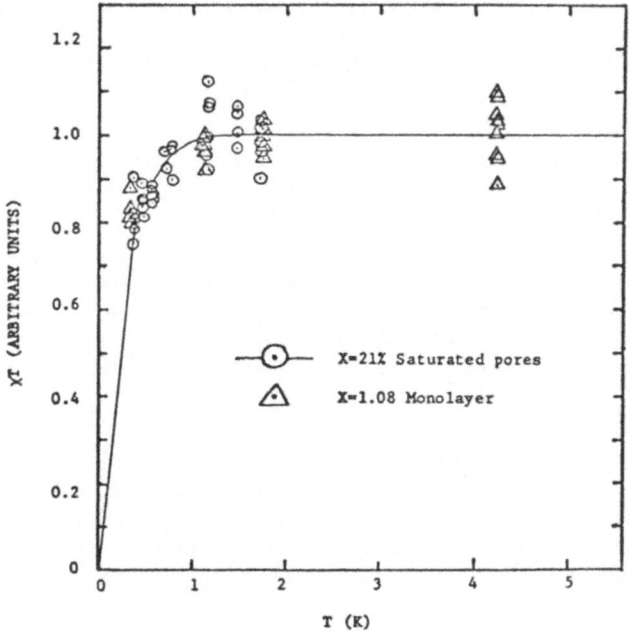

Figure 4. Product of nuclear magnetic susceptibility and temperature
          vs. temperature for various coverages (x) of He³ adsorbed
          on Grafoil.

is due to liquid-like behavior of the thick layer. However, for
the monolayer and submonolayer cases it is not entirely clear
whether the similar fall-off in susceptibility is due to statistical effects or due to some other causes. Measurement of χ from
FID is complicated by the fact that the initial part of the decay
component in the FID is unobservable within the receiver recovery
time ($\sim 35 \mu s$). That is, growth of a signal component with T_2
much shorter than 35 μs, caused possibly by increased substrate-
He^3 interaction at the lower temperatures, would lead to an apparent fall-off in susceptibility.

We attempted to measure spin-diffusion using a 90°-τ-180°
pulse sequence, and a linear gradient of up to 3 gauss/cm. The
standard theory of spin-diffusion[5] predicts that the logarithm of
the echo-envelope will go as τ^3. Experimentally the 90°-180°-
echo envelope decays were found to be exponential ($e^{-\tau/T_2}$)
and independent of gradient for the monolayer coverages. However,
for the multilayer coverages greater than x=2 ML the logarithm of
the 90°-180°-echo envelope was found to be gradient dependent although not in a τ^3 manner. These data can be explained by the
fact that the adsorbed He^3 is bounded by the finite size of the
graphite flakes comprising the Grafoil sheet.

Room temperature NMR data were taken on F^{19} in liquid hexafluorabenzene (C_6F_6), both in bulk and saturating Grafoil samples.
These room temperature experiments were done to gain a better understanding of the electrical and magnetic effects of Grafoil on
an NMR signal, obtained with a spin system with well defined and
known characteristics in a bulk liquid sample. At room temperature C_6F_6 sample taken from the bottle without further purifica-
tion has a $T_1 \sim 2T_2$ of the order of 1.0s, and the coefficient of
self-diffusion (D) in the range of 1.0×10^{-5} cm^2/s.

In addition, in an attempt to separate the effects associated
with adsorption from the effects of finite electrical conductivity
of the substrate, NMR measurements were done on C_6F_6 saturating
porous paper, and in bulk C_6F_6 surrounded by aluminum and Grafoil
shields.

In summary, the following can be said about these experiments:
(1) Grafoil tape, separated by teflon sheet, wrapped around a <u>bulk</u>
liquid sample does not affect T_1 or T_2 and has no observable
shielding effects. However, the FID time is reduced by about a
factor of two when Grafoil is placed around the sample, that is,
the uniformity of the external magnetic field over the sample is
affected by Grafoil placed in its vicinity.
(2) C_6F_6 saturating insulated rolled Grafoil, an arrangement similar to the one used for He^3 experiments, has a dramatically
shortened FID ($\sim 200 \mu s$) and T_2(~ 2.5 ms), while T_1 is only

slightly less than in bulk liquid. There are no diffusion effects observable when an external gradient is applied. For a laminated substrate with Grafoil planes parallel to rf field H_1, the FID and T_2-decay are dependent on the orientation of the substrate in the external magnetic field H_0.

(3) C_6F_6 saturating fine-pored paper (with pores $\sim 100\mu$ in size) has the same FID time as the bulk, a T_1 slightly less than the bulk, but a much shorter T_2 (~ 45 ms). An external gradient has no effect on the apparent T_2. When large-pored blotter paper (with pores $\sim 400\mu$ in size) was used as a sponge, T_1 was the same as the bulk liquid sample, with a T_2 of about 90 ms, and external gradient effects were observable. Calculations using the Robertson theory[7] of bounded diffusion show that in both cases pore size effects would override gradient effects, in agreement with experimental observations.

The room temperature NMR on F^{19} in C_6F_6 thus produces some of the effects observed for adsorbed He³ : shortened FID's and spin-echo envelopes independent of the field gradient. Since self-diffusion in bulk C_6F_6 is large ($D \sim 10^{-5}$ cm²/s) and readily observable, this means that even in saturated C_6F_6 films the effects of bounded diffusion play a dominant role in determining the dependence of the echo envelope on the field gradient. Also, the magnetic properties of the Grafoil play a dominant role in the shortening of the FID and T_2-decay.

Interpretation of the NMR data for He³ adsorbed on Grafoil is clearly rather difficult, because the substrate plays the main role in determining the properties of the adsorbed He³ spin system and its magnetic environment. Magnetic interactions between He³ atoms are in effect masked by the very much stronger interactions between He³ and the electrically conducting substrate (Grafoil). The observed T_2-decay is believed to be determined mainly by the spin interaction between He³ and electrons in the substrate, which is much larger than the He³-He³ interaction. The He³-He³ interaction would, as a rule, be sensitive to any order transitions which have been suggested by thermodynamic measurements[8]; however, our T_2 data on Grafoil do not reflect any such phase transitions in the He³ system. Direct measurements of He³ spin-diffusion in the external magnetic field gradient are not possible because of physical constraints imposed by the Grafoil structure. The He³ spin-lattice interaction is clearly also dominated by the substrate. The scatter in the observed T_1 values is very large and it is difficult to assign specific behavior to T_1. Magnetic susceptibility of the He³ spin system, in principle, offers a test of the thermodynamic state of the system. The measured susceptibility of adsorbed He³ falls below that given by Curie's law as the temperature of the system is lowered below 1 K. This result can be fitted to an ideal Fermi gas behavior with a degeneracy temperature in the range of 0.5 K. However, the observed susceptibility could also be explained by evolution of an undetected short T_2 component in the nuclear signal which grows with decreasing temperature; there is yet no evidence that such a short component develops in fact as the temperature is lowered. Also, the observed behavior of the magnetic susceptibility can possibly be explained in terms of the relatively strong He^3-substrate interaction.

Because of the numerous complications which arise in the interpretation of NMR data for Grafoil substrate, we have recently performed pulsed NMR measurements on Sterling FT pyrolytic graphite[2] which like Grafoil has a uniform surface.[9] Also, this substrate has been used in recent continuous wave (CW) NMR measurements by Rollefson[10], who reported a marked anomaly in the vincinity of 0.6 ML.

Our Sterling FT sample was carefully prepared closely following the procedure reported by Rollefson.[10] The Sterling FT was baked in a quartz tube in a vacuum furnace at $1100^{\circ}C$ for over 48 hours. The baking tube was then placed in a helium purged glove box, the tube opened, and a quantity of powder weighed. The sample chamber was assembled and placed onto the low temperature probe, all done in a helium atmosphere. The manufacturer's data for specific area ($13m^2$/gram) was used to determine a He^3 monolayer capacity.

Measurements of the FID, T_2 and T_1 were made at submonolayer coverages for $T \sim 4.2K$, 2.92K, and 1.75K. Equilibrium of the system was determined by monitoring susceptibility and FID time. Typical time to reach equilibrium was about 8 hours. At 4.2K the system came to equilibrium much faster than at 1.75K, presumably because of the non-negligible vapor pressure above the adsorbed system.

The following can be said about these data:
(1) The FID (T_2') and T_2-decay times at all coverages increases with increasing temperature. At 4.2K the T_2 ($\sim$20ms) $\gg T_2'$ ($\sim$2.5ms), but at 1.75K $T_2 \sim T_2'$ ($\sim$1ms). For He^3 adsorbed on Grafoil, $T_2 \sim 10\ T_2'$ for $x \sim 1$ ML and T=4.2K and 1.16K.
(2) T_1 is longer at 4.2K ($\sim$1.25 sec) than at 1.75K ($\sim$0.5 sec) except for a monolayer coverage where the T_1 at 4.2K is the same as at 1.75K ($\sim$0.5 sec). For He^3 adsorbed on Grafoil, T_1 increases with decreasing temperature.
(3) The signal to noise ratio is considerably higher for Sterling FT, as compared to Grafoil, for the same number of He^3 spins. This is presumably due to the decreased shielding with Sterling FT.
(4) There is a spin-echo broadening effect similar to what is seen with Grafoil.

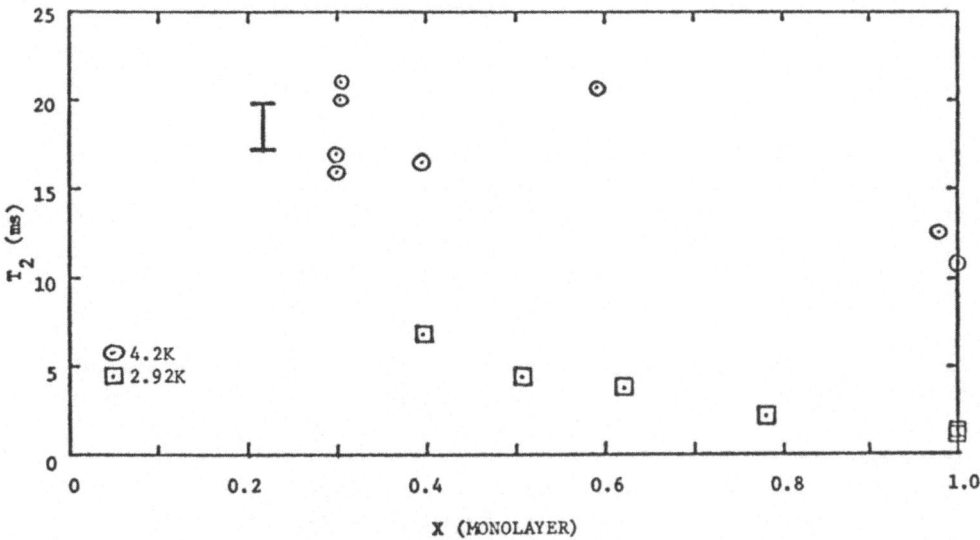

Figure 5. Spin-spin relaxation time (T_2) as a function of temperature and coverage for He³ adsorbed on Sterling FT.

We are now in the process of making additional measurements on this system and analyzing the data. Our results show that there is an apparent discrepancy between Rollefson's CW NMR line-width measurements, and our pulsed NMR measurements. Since $T_2' \sim [\gamma \cdot \delta H]^{-1} \sim 50 [\delta H (gauss)]^{-1} \mu s$ for He³, where γ is the nuclear gyromagnetic ratio and δH is the NMR linewidth, the linewidth of ~ 1.5 gauss for 0.9 ML at 1.75K as reported by Rollefson, corresponds to a T_2' $\sim 33 \mu s$. The corresponding FID time measured by us at 1.0 ML is of the order of $400 \mu s$. Also, under these conditions we find T_2 to be the same as the FID time. In addition, we have performed a series of measurements at a constant temperature with gradually increasing coverage in an attempt to observe any anomalies which might occur as the coverage is increased. At 2.92K we find that T_2 decreases monotonically from ~ 6.8ms at x=0.4 ML to $\sim 900 \mu s$ at x=1.0 ML. Thus, we find no evidence of any anomaly in T_2 (Fig. 5). Similarly, T_1 decreases with x in a monotonic manner from ~ 830ms at x=0.4 ML to 400ms at x=1.0 ML. We are now continuing our investigation of the Sterling FT system. The results so far obtained indicate that, as in Grafoil, the magnetic properties of the He³ system adsorbed on Sterling FT are dominated by the He³-substrate interaction which effectively masks any changes in the much weaker He³-He³ interaction.

REFERENCES

*Work supported in part by National Science Grant No. GU-1147, GP-24572 and Navy Equipment Loan Contract No. NONR-1758(00).

(1) The Grafoil sheet, GTA grade, was obtained from Union Carbide, Carbon Products Division, Cleveland, Ohio.
(2) Sterling FT (similar to graphitized P-33) was obtained from the Cabot Corporation, Boston, Massachusetts. These graphite particles are $\sim 0.2 \mu m$ in diameter and are polyhedra with the basal plane exposed. The specific area is given as $13 m^2$/gram of powder.
(3) D. P. Grimmer and K. Luszczynski, "NMR in He^3 Monolayers Adsorbed on Graphite Below 4.2K," Proceedings of the Thirteenth International Conference on Low Temperature Physics, Boulder, Colorado, August 1972.
(4) M. Bretz and J. G. Dash, Phys. Rev. Letters 26, 963 (1971).
(5) H. Y. Carr and E. M. Purcell, Phys. Rev. 94, 630 (1954).
(6) The hexafluorabenzene (C_6F_6), 99% purity, was obtained from the Aldrich Chemical Co., Inc., Milwaukee, Wisconsin.
(7) B. Robertson, Phys. Rev. 151, 273 (1966).
(8) M. Bretz and J. G. Dash, Phys. Rev. Lett. 26, 963 (1971); M. Bretz and J. G. Dash, Phys. Rev. Lett. 27, 647 (1971); D. C. Hickernell et. al., Phys. Rev. Lett. 28, 789 (1972).
(9) A. Thomy and X. Duval, J. Chim. Phys. Physico-Chim. Biol. 66, 1966 (1969).
(10) R. J. Rollefson, Phys. Rev. Letters 29, 410 (1972).

PULSED NMR MEASUREMENTS OF ETHYL ALCOHOL ABSORBED IN GRAFOIL

D. L. Husa and D. C. Hickernell
Cryogenics Center
Stevens Institute of Technology, Hoboken, NJ

J. E. Piott
Yeshiva University, New York, NY

INTRODUCTION

Previous experiments[1] have shown that the large anisotropic diamagnetism[2] of graphite determines the observed NMR linewidth of a fluid absorbed in Grafoil[3,4]. This linewidth intrinsic to liquids in Grafoil ($\Delta H \sim$ 0.25 gauss for a magnetic field of 7.05 Kgauss perpendicular to the Grafoil sheets) could be responsible for large magnetic field gradients over the small voids in the Grafoil. In that event pulsed NMR spin echo measurements of any fluid absorbed in Grafoil would exhibit the effects of bounded diffusion in a magnetic field gradient[5,6]. In order to test that hypothesis a series of spin echo measurements were made on a fluid of known coefficient of diffusion absorbed in Grafoil at room temperature. The results show factors that must be considered in interpreting NMR measurements on He^3 in Grafoil.

EXPERIMENTAL PROCEDURE

Pulsed NMR measurements were carried out at 15, 30 and 40 MHZ with a high power fast recovery apparatus similar to those used elsewhere[7-9]. Since no preamplifier was used before the superheterodyne receiver at 15 MHZ and 40 MHZ, the signal to noise ratio was poorer than at 30 MHZ. Grafoil sheets 0.013 cm. thick were separated by mylar sheets 0.0026 cm. thick and then formed into a cylindrically shaped sample as previously

described[1]. The sample (about 1.2 cm. in diameter and
7.5 cm. long) was saturated with ethyl alcohol denatured
with about 1% hydrocarbons. This should decrease D from
the pure alcohol value of (1.0) (10^{-5}) cm^2/sec by a
small amount and not seriously affect the bounded diffusion.

Since the r.f. coil was about 1.2 cm. long, the r.f.
field H_1 of the 90° and 180° pulses was not homogeneous
over the 7.5 cm. long sample. While measuring the transverse relaxation time T_2 by a 90° - 180° spin echo sequence in the presence of such a H_1 inhomogenity, care
must be taken to avoid placing the 180° pulse in the
free induction decay (FID) of the 90° pulse[10]. With a
FID lasting approximately 500 μsec, the amplitude of the
spin echo could not be measured for times less than one
millisec. In order to measure the spin echo envelope for
times as short as 100 μsec, the sample was placed in an
inhomogeneous point in the magnet where the applied magnetic field had an inhomogenity of about 5 gauss/cm.
Both of the resulting spin echo decay envelopes are
shown in Figure 1.

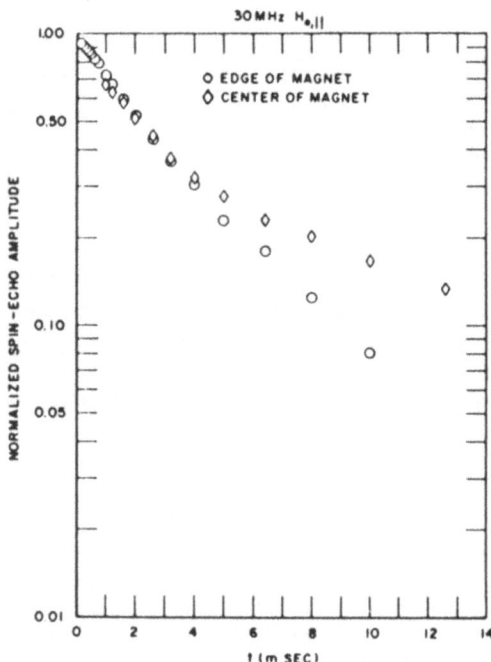

Fig. 1 Amplitude of the spin echo as a function of the
time t from the 90° pulse to the echo: ◇ center of field
○ fringefield of the magnet(magnet gradient≈5 gauss/cm.)
The magnetic field is 7.05Kg(f_0=30MHZ) and parallel to
the Grafoil sheets.

Note that the initial part of the decay remains unchanged. The non-exponential decay observed at long times for the sample in the center of the field may be due to the percentage of the alcohol that is between the grafoil sheets or a distribution of void sizes. Since a magnetic field gradient of about 5 gauss/cm. increases the decay rate at long times, it appears that the alcohol between the sheets is the primary cause of the non-exponential decay. Bounded diffusion in a large gradient (about 1000 gauss/cm) would be unaffected by an additional gradient of 5 gauss/cm.

The results reported here were carried out on Grafoil that had been heated to drive off paramagnetic impurities (final concentration < 100 ppm.) As a test, spin echo measurements were carried out on uncleaned 0.076 cm. thick Grafoil. Since the T_2 (spin echo) for the Grafoil 6 times thicker was only 15% lower, we believe our T_2 measurements on the clean thin Grafoil are unaffected by r.f. shielding or paramagnetic impurities.

RESULTS AND DISCUSSION

Figure 2 shows the spin echo decay envelopes for the magnetic field H_o parallel to the sheets of the Grafoil. Note that the decay is not t^3 as for unbounded diffusion in a magnetic field gradient. Moreover, the roughly exponential decay is dependent on the applied magnetic field and is much faster than the intrinsic transverse relaxation T_2 of ethyl alcohol.

If one postulates that the linewidth $(\Delta H)_{11} = (0.15)$ gauss observed at 30 MHZ[1] for H_o parallel to the sheets is entirely due to the variation of the magnetic field across the length of each void, the gradient in a void $a = 1.3\mu$ long is

$$G_{11} = \frac{(\Delta H)_{11}}{a} \simeq 1,150 \text{ gauss/cm.} \tag{1}$$

where G_{11} is magnetic field gradient along the length of the void for $H_o = 7$ Kg parallel to the Grafoil sheets. Although the total variation in magnetic field is small the gradient is very large because of the small void size.

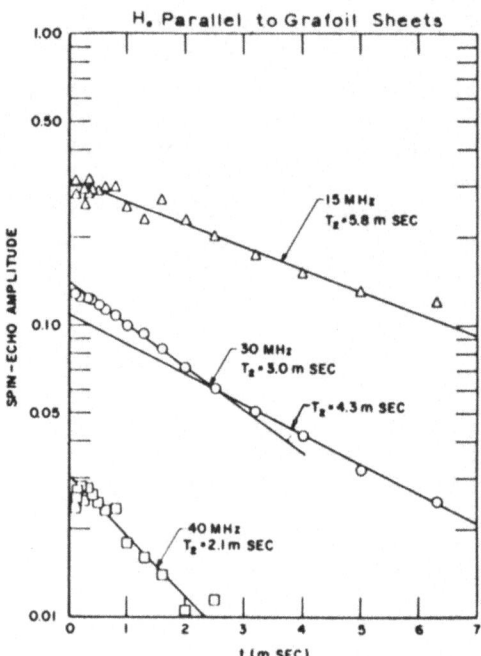

Fig. 2 Amplitude of spin echo as a function of t for three values of the magnetic field. The magnetic field is parallel to the sheets.

Including the effect of wall relaxation, the transverse relaxation time is given by

$$\left(\frac{1}{T_2}\right)_{\substack{11 \\ observed}} = \frac{1}{T_{2w}} + \frac{a^4\ \gamma^2\ G_{11}^2}{120\ D} \tag{2}$$

where $a^4\ \gamma^2\ G_{11}^2/120D$ is the asymptotic form of Robertson's formula[6] for bounded diffusion, γ is the gyromagnetic ratio for protons, and T_{2w} is the wall relaxation. We have assumed the wall relaxation is exponential. Let us define σ_{11} by

$$G_{11} = \frac{(\Delta H)_{11}}{a} = \frac{\sigma_{11}\ H_o}{a} \tag{3}$$

where $\sigma_{11} = [\Delta H_{11}(T)/H_o]$ is a constant for any particular temperature. The fact that σ_{11} is independent of H_o (ie. $\Delta H \alpha H_o$) has been established previously[1]. The temperature dependence of $\Delta H_{11}(T)$ reflects the temperature dependence of the diamagnetism of graphite.

Equation (2) now becomes

$$\left(\frac{1}{T_2}\right)_{11}^{observed} = \frac{1}{T_{2w}} + \left(\frac{a^2 \gamma^2 \sigma_{11}^2}{120\ D}\right) H_o^2 \qquad (4)$$

where the coefficient of H_o^2 is independent of the magnetic field if the temperature is held constant. The inverse of the spin echo decay constant for H_o perpendicular and parallel to the sheets is plotted vs H_o^2 in Fig. 3.

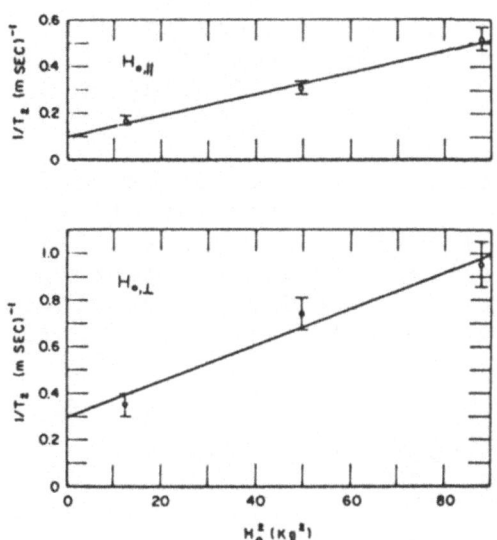

Fig. 3 The inverse of the slope of the spin echo decay curve for t between 0.5 and 2 msec. is plotted vs. H_o^2. The slope of <u>this</u> graph is ($a^2 \gamma^2 \sigma^2/120\ D$) and has the value (4.6) (10^{-6}) gauss^{-2} sec^{-1} for H_o parallel to the sheets and (7.78) (10^{-6}) gauss^{-2} sec^{-1} for H_o perpendicular to the sheets.

The intercepts of the straight lines gives the wall relaxation. In view of the scatter in the data, the values of T_{2w} derived from those intercepts are not very accurate. However, the slopes are definitely not the same and reflect the difference between σ_{11} and $\sigma_{\perp}$ = $(\Delta H)_{\perp}/H_o$ where $(\Delta H)_{\perp}$ is the linewidth when the sheets are perpendicular to H_o. Using the experimental slopes one can then calculate what "average" value of a gives the observed decay. Denoting by a_{11} the value determined from the slope for H_O parallel to the sheets and defining a similarly for the perpendicular case:

$$a_{11} = 1.30\mu \qquad (5)$$

$$a_\perp = 1.01\mu \qquad (6)$$

Those values compare favorably with the values of 1-4 μ observed by microscope[3]. In view of the complex nature of the geometry of the voids in Grafoil, one should not expect precise values for the magnetic field gradient or void size in Grafoil. It should be pointed out that spins diffusing from void to void would give the same type of spin echo decay as above. The Robertson analysis was developed for bounded diffusion but would also apply to a spin diffusing in a periodic magnetic field.

The asymptotic form used for bounded diffusion in equation (2) is valid only for long times, that is $t \gg (2/\pi^2)(a^2/D)$. For $a = 1$ μ and $D = (10^{-5})$ cm^2/sec, the condition is $t \gg 0.2$ msec. Since we took the slope of the decay curves for t from 0.5 to 2 millisec, that condition is roughly satisfied as shown in Figures 4 and 5.

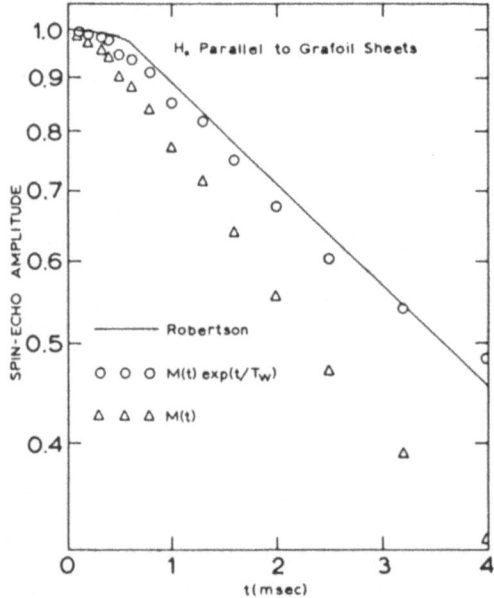

Fig. 4 Robertson's theory is compared to the spin echo data(f_o=30MHZ): $\Delta M(t)$; O $M(t)$ $e(t/T_{w11})$ where T_{2w11} is 10 msec and H_o is parallel to the sheets. The theoretical curve corresponds to a = 1.3μ, $(\Delta H)_{11}$ = 0.15 gauss, G_{11}= 1,150 gauss/cm and has (T_2) asmptotic = 4.4 msec.

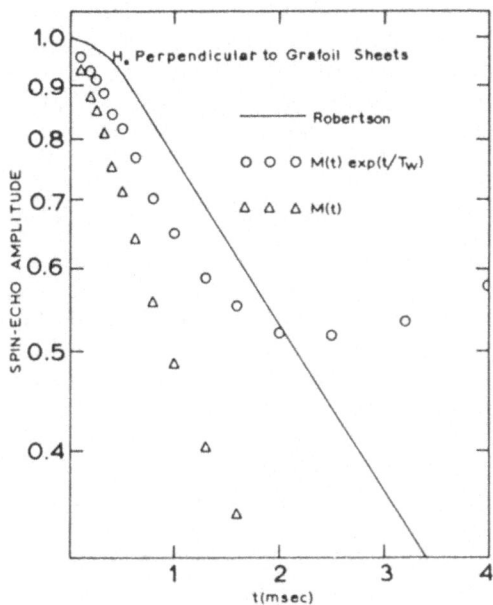

<u>Fig. 5</u> Robertson's theory is compared to the spin echo data (f_o = 30 MHZ): Δ M(t) ; $\bigcirc$ M(t) $_e$(t/$T_{2w\perp}$)where $T_{2w\perp}$ = 3.45 msec and H_o is perpendicular to the sheets. The theoretical curve corresponds to a = 1.0μ, ($\Delta H)_\perp$ = 0.25 gauss, $G_\perp$ = 2,500 gauss/cm and has (T_2) asymptotic = 2.7 msec.

Note that the theoretical decay curve starts out slowly (a t^3 dependence) and then becomes exponential as the boundedness of the diffusion becomes important. The experimental points have been multiplied by $e^{(t/t_{2w})}$ to correct for wall relaxation. The agreement is quite good for H_o parallel to the sheets but not for H_o perpendicular to the sheets. This may be due to the fact that bounded diffusion relaxation for H_o perpendicular to the sheets proceeds faster and hence any alcohol between the sheets becomes more apparent at short times. Furthermore, the estimated wall relaxation rate is larger for H_o perpendicular to the sheets and any significant error in it can seriously affect the curve. A non-exponential wall relaxation may also be present as will be noted shortly.

Given the density of grafoil and graphite, the surface area of grafoil per gram, and the fact that the voids are very flat, one can compute the thickness b of the voids. It is about 0.04μ. That is comparable to the 0.05 to 0.1μ platelet thickness observed by microscope[3]. If one notes that the Robertson equation also applies to diffusion across thickness of the void, eqn. (4) can be rewritten as

$$\frac{1}{T_{2_{\parallel}}\text{observed}} = \frac{1}{T_{2w}} + \frac{a^2 \gamma^2 \sigma_{11}^2 H_o^2}{120\ D} + \frac{b^2 \gamma^2 \sigma_{11}^2 H_o^2}{120\ D}$$

where we have assumed that the gradient across the dimension b is given by $G = (\Delta H)_{11}/b = \sigma_{11} H_o/b$. Since $a \sim 1\mu$, the contribution to the decay by the "b term" is much smaller than that from the "a term."

The effects of surface relaxation are also important. Wayne and Cotts[5] have observed that surface relaxation occurs for methane gas in small cavities and attribute this to physical adsorption of the gas molecules. In other words a surface layer of "fluid" behaves as if it were a rigid or semirigid lattice. The motional narrowing is thus decreased[11] and T_2 is much shorter. They also found that this increased transverse relaxation rate, as measured by the Carr-Purcell[12] Method B ($90° - 180° - 180° - 180° \ldots$) was not exponential but was greater at short times than for long times.

Similar effects are probably present when ethyl alcohol is absorbed in Grafoil. They have been treated theoretically by Robertson[6]. Since none of these effects depend upon the strength of the applied magnetic field, they should not affect our analysis of the magnetic field dependence of T_2 in terms of bounded diffusion. However, if the surface layer has a chemical shift due to the graphite surface, then $1/T_2$ has a term proportional to H_o^2 in the presence of fast nuclear transfer between the surface and bulk phases. Note that this is the same functional form as bounded diffusion.

Woessner[13] has derived expressions for T_2 in the presence of nuclear transfers between two phases with a chemical shift, never seen on films. However, in order to calculate the magnitude of T_2 one needs to know the chemical shift, transfer rate, and fraction of spins in each phase. Since we know none of these quantities, we can only estimate the importance of this mechanism compared to bounded diffusion. This estimate has been done and it was found that unless the chemical shift for monolayers on Grafoil is hundreds of times larger than for silica gel[14] the effect is negligible.

An experimental measurement that differentiates between the two possible mechanisms is needed. One such
technique is $T_{1\rho}$, the relaxation rate of magnetization
along a rotating magnetic field[15]. Since the Grafoil
field inhomogenity (at 30MHZ) is about 0.25 gauss, a
rotating field of about 10 gauss should eliminate dephasing due to the field inhomogenity[16]. Thus one
could distinguish between bounded diffusion (a result
of diffusion in the presence of magnetic field inhomogenity) and exchanges between phases of different
chemical shifts[17]. Rotating reference frame experiments to decide this point are now in progress.

CONCLUSIONS

The relaxation rate in a spin echo experiment on
alcohol in Grafoil has been found to contain a term
proportional to H_o^2. This dependence on H_o^2 was attributed to the bounded diffusion of spins in a large magnetic field gradient present in the Grafoil voids.
Analysis on such a basis yielded a void size in agreement with that observed by microscope.

Provided a thin film such as He^3 experiences the
same effective magnetic field gradients as a fluid completely filling the Grafoil voids, one can then invert
the analysis and use the parameters determined in this
experiment to obtain a coefficient of diffusion for
the film. However, before such an analysis is attempted,
the experiments presented here should be repeated for
a fluid with a known coefficient of diffusion significantly different from that of alcohol. In any case,
the same effects studied here are most likely present
in monolayer films. Any NMR observations should be
analyzed from this viewpoint.

REFERENCES

1. D. C. Hickernell, D. L. Husa, J. G. Daunt, and
    J. E. Piott, to be published.

2. J. W. McClure, Phys. Rev. 119, 606 (1960)

3. D. P. Grimmer and K. Luszczynski, Proceedings of
    the 13th International Low Temperature Conference,
    Boulder, Colorado (1972) to be published.

4. R. J. Rollefson, Phys. Rev. Lett 29, 410 (1972)

5. R. C. Wayne and R. M. Cotts, Phys. Rev. 151, 262
    (1966)

6. Baldwin Robertson, Phys. Rev. 151, P. 273 (1966)

7. I. J. Lowe and C. E. Tarr, J. of Sci. Inst. 1,
    P. 320 (1968)

8. I. J. Lowe and C. E. Tarr, J. of Sci. Inst. 1,
    P. 604 (1968)

9. J. D. Ellett, Jr., M. G. Gibby, U. Haeberlen,
    L. M. Huber, M. Mehring, A. Pines and J. S. Waugh,
    Adv. in Mag. Res. 5, edited by J. S. Waugh, Aca-
    demic Press, New York, (1971) P. 153

10. Anil Kumer and Charles S. Johnson, Jr., J.of Mag.
    Res. 7, P. 55 (1972). This article treats T_1
    measurements in the presence of an inhomogeneous
    H_1. Similar considerations apply for T_2 measure-
    ments.

11. N. Bloembergen, E. M. Purcell and R. V. Pound,
    Phys. Rev. 73, P. 679 (1948)

12. H. Y. Carr and E. M. Purcell, Phys. Rev. 94, P. 630
    (1954)

13. D. E. Woessner, J. Chem. Phys. 35, P. 41 (1961)

14. D. E. Woessner and J. R. Zimmerman, J. Phys. Chem.
    67, P. 1590 (1963)

15. D. C. Look and I. J. Lowe, J. Chem. Phys. 44, P. 2995
    (1966)

16.   Thomas C. Farrar and Edwin D. Becker, Pulse and Fourier Transform NMR, (Academic Press, New York 1971) P. 91

17. C. Deverell, R. E. Morgan and J. G. Strange Molecular Physics 18, P. 553 (1970)

DETERMINATION OF THE ATOMIC GEOMETRY OF SUBMONOLAYER FILMS

C. B. Duke

Xerox Research Laboratories

Webster, New York, 14580

It has been known for many years that the deposition of sub-monolayer coverages of an atomic or molecular species on single crystal substrates often leads to ordered surface structures.[1-4] Such structures are characterized by periodic translational symmetry parallel to the surface of the single-crystal substrate. If the Bravis net of the combined adsorbate-substrate system is of lower symmetry than that of the substrate alone, it can be determined directly from the geometry of beams of electrons reflected from the adsorbate-covered surface. If the electrons are incident nearly normal to the surface and their energies occur in the range $10eV \leq E \leq 1000eV$, the electron-reflection process is referred to as "low-energy electron diffraction" abbreviated "LEED". If the electrons are incident at glancing angles to the surface, and their energies are greater than 1000eV, this process is called "reflection high-energy electron diffraction" or "RHEED". Descriptions of the experimental techniques for performing these measurements may be found in the review literature[1-3].

An important aspect of both LEED and RHEED is their sensitivity to the structure of the uppermost few atomic layers of the solid. The primary origin of this fact resides in the large ($\sigma \sim 10^{-16} cm^2$) inelastic electron-solid cross sections associated with the excitation of plasma oscillations ($\Delta E \sim 10eV$) in the solid by electrons in the LEED energy range[5]. Because of these large inelastic cross-sections, electrons which are elastically reflected from the solid and whose energies normal to it satisfy $10eV \leq E \leq 300eV$ must have emanated from within about 10Å of the surface.

In addition to the desirable consequence of surface-sensitivity,

the strong electron-solid interactions in the LEED energy region
have the undesirable one that the Born approximation (even in dis-
torted-wave format) fails to describe the elastic scattering cross-
sections. Therefore a multiple-scattering or "dynamical" model is
required in order to interpret observed LEED intensities. For-
tunately, however, within the past three years such models have
been constructed and shown to provide excellent descriptions of
measured intensities associated with the low-index faces of clean
metal substrates[4-6]. Thus, the stage is set for the application
of LEED to determine the atomic geometry of ordered submonolayer
adsorbed films.

The essential difficulty in obtaining the atomic geometry of
submonolayer films resides in the two facts that the contents as
well as structures of the unit cell are unknown and their determination requires an analysis of the intensities of the diffracted
electrons. Whereas the geometry of the diffracted beams is a
measure of the (two-dimensional) Bravais net of the overlayer-sub-
strate system, only a thorough interpretation of the intensity of
these beams as a function of the incident electrons' energy and
angle suffices to discern the structure of the basis associated
with this net. Moreover, the unit cells extend throughout the
sample normal to the surface being examined. Consequently both a
sufficiently accurate description of the electron-solid interactions
and a satisfactory dynamical theory of the scattering process are
necessary prerequisites for a determination of surface atomic geometries. In several recent reviews[4,6] the adequacy of these prerequisites is examined critically. The conclusion reached in these
reviews is that within quantitatively-stated limitations, these re-
quirements have become satisfied, for the first time, during the
past two years.

Two approaches have been proposed for use in surface crystal-
lography. In data-reduction methods[4,7,8] experimental LEED intensities are processed using signal-averaging techniques. Following
this, the resulting averages are analyzed using (explicitly or im-
plicitly) the Born approximation. In microscopic model
methods[4,9-13] the structure analysis proceeds by direct comparison
of observed and dynamical-model LEED intensities. Thus far, the
structures of clean surfaces[4,6,7,8] and of Rh(100)-c(2 x 8)-o[7]
have been examined by the former method and those of clean sur-
faces[4,6,9] and of Ni(100)-c(2 x 2)-Na[10], Ag(111)-($\sqrt{3}$ x$\sqrt{3}$)30°-I[11],
Ni(100)-c(2 x 2)-S[12,13] and Ni(100)-c(2 x 2)-o[13] by the latter.
Both techniques are directly relevant for the determination of the
atomic geometry of helium monolayers on single crystal substrates. Sufficiently low current densities must be used, however,
so that the He is not desorbed by the electron beam. Such experiments
have not yet been performed primarily because the design of ultra-
high-vacuum cryogenic crystal manipulators is in a rudimentary state.

Finally, since order-disorder transitions are of considerable interest in the study of submonolayer He films, we recall that in principle LEED permits a direct observation of the structure of both phases[4]. Indeed, the construction of dynamical models of the LEED process from disordered overlayers on single-crystal substrates is one of the current frontiers in LEED theory[14]. At the present time, however, only the structures of phases exhibiting long-range periodic order have been ascertained by analyses of observed LEED intensities.

References:

1. J. J. Lander, in Progress in Solid State Chemistry, Vol. 2 (Pergamon, New York, 1965), pp. 26-118.

2. J. W. May, Adv. Cataly. 21, 151 (1970).

3. C. B. Duke and R. L. Park, Physics Today 25, 23 (August, 1972).

4. C. B. Duke, Adv. Chem. Phys. (in press).

5. C. B. Duke and C. W. Tucker, Jr., Surface Sci. 15, 231 (1969); Phys. Rev. Letters 23, 1163 (1969).

6. C. B. Duke, in LEED: Surface Structure of Solids, Vol 2, edited by L. Laznicka (Union of Czechoslovak Mathematicians and Physicists, Prague, 1972), pp. 125-291.

7. C. W. Tucker, Jr. and C. B. Duke, Surface Sci. 29, 237 (1972).

8. T. C. Ngoc, M. G. Lagally and M. B. Webb, Surface Sci. 35, 117 (1973).

9. G. E. Laramore and C. B. Duke, Phys. Rev. B 5, 267 (1972).

10. S. Andersson and J. B. Pendry, J. Phys. C 6, 601 (1973).

11. F. Forstman, W. Berndt and P. Büttner, Phys. Rev. Letters 30, 17 (1973).

12. C. B. Duke, N. O. Lipari, G. E. Laramore and J. B. Theeten, Solid State Commun. (in press).

13. C. B. Duke in Proceedings of the International School of Physics Enrico Fermi, Course LVIII (Academic Press, New York, 1973).

14. C. B. Duke and A. Liebsch, Bull. Am. Phys. Soc. 18, 364 (1973); Phys. Rev. (submitted).

STRUCTURE SENSITIVE SCATTERING OF ATOMS AND MOLECULES FROM SOLID SURFACES

H. Saltsburg

University of Rochester

Rochester, New York   14627

Introduction

The study of the collisional process in the initial interaction between a gas particle and solid surface is one means of investigating the structure of surfaces. The collisional interaction must not, however, modify that structure but simply reflect it implying that elastic interactions should be of primary interest. Inelastic interactions (including reactive interactions) are of lesser utility in structural studies. In the following a brief review of the use of atomic and molecular beam scattering for surface structure determination is presented (1).

Experimental Techniques

The typical experimental apparatus has been described in detail elsewhere (2). It suffices to note that one needs to translate into an experimental device the idealized picture of a single particle approaching a surface, interacting with that surface and either sticking to or leaving the surface. The process is observed typically by means of those particles which leave the surface. The apparatus consists then of five major components:

1. A source of particles which act independently in the collision event (typically a Knudsen effusion (3) or nozzle beam source (4)).

2. A well-defined surface prepared as carefully as pos-
    sible with regard to structure and composition.

3. A detector to observe the results of the scattering
    process in terms of spatial, momentum, or energy dis-
    tributions of the scattered beam.

4. A set of collimators to define and distinguish be-
    tween the incident and scattered beams.

5. A vacuum system capable of reducing the background
    pressure to eliminate extraneous gas phase collisions
    and insure that only the initial collision of particle
    and surface is being observed. (Electronic aids,
    utilizing lock-in detection, typically are used to
    distinguish primary and secondary collision events
    (5).)

The typical measurement is that of the spatial distribution of scattered particles. Usually one does not
measure the energy of the scattered beam but rather one
infers from the spatial distribution the energetics of
the collisional process (6). Specifically, specular
reflection (characterized by a peak in the specular angle
and a half-width equal to that of the appropriate diver-
gence of the incident beam) implies an elastic collision
with no change in the normal momentum. If the deBroglie
wavelength of the particle is appropriate, diffraction
may also result. Complete thermal accommodation is
usually characterized by diffuse or cosine-law scattering.
Intermediate cases are characteristic of inelastic processes involving the dynamical properties of the scattering system. In all cases, structural artifacts may alter
the scattering distributions (6).

Scattering of He Atomic Beams

The earliest molecular-beam surface-scattering
experiment was a study of He scattered from LiF (7).
Diffraction was observed and it was found that the appropriate lattice for diffraction had the characteristic
spacing of the two dimensional like-ion grid of the alkali
halide (Observation of H_2 and H atom scattering from
LiF and other alkali halides confirmed the existence of
diffraction (8).). Other studies showed that both ^{3}He
and ^{4}He exhibited intense diffraction yet D_2 showed
little diffraction. Clearly, molecular structure plays
a role in determining whether elastic behavior is exhi-

bited (9) (10).

In contrast, clean (111) surfaces of FCC metals
exhibit no diffraction of He . Only the (110) azimuth
of (112) W exhibits diffraction among all clean metal
surfaces investigated (11). It appears likely that the
scattering potential does not vary sufficiently rapidly
in the surface plane for He diffraction to be observed
as a dominant spatial characteristic from most metal
surfaces (12). These surfaces may not only be too smooth
but may permit significant inelastic scattering as well.
LEED observations of surface structure depends upon distinguishing the very small fraction of elastic events
from the much larger fraction of inelastic events (13).
A purely spatial separation is not found.

A carbonized W(110) surface has been shown to dif-
fract He from a structure which is consistent with that
observed as an overlayer structure using LEED techniques
(14). If inelastic attenuation due to thermal vibrations
of the lattice (a Debye-Waller effect) is a dominant loss
mechanism, the carbide structure must have a very high
Debye temperature relative to metals. Scattering of He
from diamond however does not show diffraction events
(15). This may be the result of a disordered surface
roughness in contrast to the ordered structure which
causes He diffraction from W(112).

The geometric symmetry of the structure of clean
surfaces can be deduced therefore in some cases but not
in all and one cannot yet predict which surfaces will be
amenable to study since the spatial distributions are
subject to significant modification by topography and
impurity. Energy resolution in the scattered beam will
aid in resolving these difficulties.

Two other effects in the scattering of He from
LiF surfaces have been noted: "selective adsorption"
and coherent inelastic scattering.

The observation of intensity minima in the angular
scan of diffraction peaks has been described by Lennard
Jones and Devonshire (16) as the result of a process in
which an incident He atom undergoes a resonant transition to a bound state resulting from the He-crystal
interaction potential. This transition occurs only under
certain momentum conditions for the incident or diffracted
beams and leads to motion of the He atom along the surface without loss of energy. Unless an inelastic col-

lision occurs to destroy the coherence, remission would
occur and minima would not be observed: the presence of
minima imply the occurence of an inelastic event. The
theory showed that bound state levels could be determined without regard to potential forms. The phenomena
is called "selective adsorption" for it occurs only under
specific momentum conditions. Both H_2 and D_2 have
been found also to exhibit selective adsorption with comparable energy levels (10). H and D atoms also show
the effect (17). Surprisingly, levels for ^3He and
^4He were found to differ by a factor of two (10). Most
other data on adsorption energies of these isotopes show
much smaller differences (18). The discrepancy is unre-
solved. It would be of interest to perform these selective adsorption experiments at much lower crystal temperatures to reduce the thermally induced attenuation and
determine if the inelastic loss is due to a collision
with structural imperfection or results from phonon
interaction.

Another type of diffraction has been reported by
Williams (19) who has shown that inelastic coherent
scattering was observable in the spatial distribution of
He scattered from LiF. Williams was able to derive
(with suitable assumptions as to the nature of the phonon
process) a dispersion curve for LiF without velocity
analysis of the scattered beam. Comparison with surface
Rayleigh wave analysis was consistent with the assumption
of the involvement of a single surface phonon.

Scattering From Adsorbed Films

The usual effect of the presence adsorbed gases on
metal surfaces is the attenuation of directed He scattering with appropriately increased diffuse scattering.
This is explained typically in terms of increased energy
transfer due to the presence of light gases adsorbed on
heavy substrates (6). Unfortunately, this is not always
the case and no general description can be given. For
example, when O_2 interacts with W the directed scattering of He is increased under some conditions and
decreased under others (20). The chemical composition
of the "LiF surface" is subject to uncertainty as there
is evidence for significant water contamination (10).
Yet, diffraction is readily observed from the outgassed
surface (under conditions where water is supposedly not
removed). Ethanol adsorbed on LiF has been reported to
exhibit extra He diffraction associated with the

absorbed film structure (21). Epitaxial Ag on LiF (22) and Hg films on LiF (23) do not show any evidence of structure when He is scattered from them. In general superlattice structures on metals with adsorbates observed by LEED techniques are not seen using He scattering. The carbon R(3x5) on W(110) is the only exception to date.

Other Rare Gases

The scattering of Ar and Ne from LiF exhibits spatial structure unlike scattering from metals. It has been described by McClure as a result of a classical elastic scattering event analagous to rainbow scattering (24). It has been observed most strikingly with Ne (25) and again reflects the structure of the scattering lattice rather than the dynamical processes of the interaction. A potential for Ne-LiF has been constructed to describe the data but is not able to be constructed via the usual simple sum of pairwise interactions (26). It is interesting to note that although many theoretical models of classical scattering from metal surfaces showed rainbow effects (27) only on LiF has it been observed experimentally.

Conclusions

The use of rare gas scattering to probe the structure of metal and alkali halide surfaces has been fruitful.

The consistent observations of diffraction and rainbow scattering from alkali halides shows the presence of a well defined geometrical structure but the intensity distributions between diffracted beams are not in agreement when different experiments are compared. This might be due to impurities which, although reflecting the halide structure are probably present as incorporated ions (e.g. H^+ and OH^- rather than adsorbed H_2O). Their presence would affect intensities rather than the location of the Bragg peaks.

The observation of selective adsorption states allows one to determine certain energy levels of the gas surface system. At extremely low temperatures these states must make significant contributions to the thermodynamic functions describing the adsorbed layer and should be observable by thermodynamic measurements. The use of alkali halides as substrates offer a well-defined system which

is amenable to other types of investigations. High area
single crystal systems can be prepared (28).

Metal surfaces are more structureless in terms of
the interaction potential. Strong screening of ion cores
is possible and only significant regular disruptions of
local structure (such as occur in high index planes) yield
diffraction effects. The development of strongly site-
ordered adsorption structures on metals is evident from
LEED studies and consequently it is likely that inelastic
events dominate the scattering from such structures.

Although diffraction from W(112) and carburized
W(110) have been observed, the probability of diffraction
is not uniform over the incident beam momentum. The diffraction peaks are narrower than the incident beam,
implying momentum selection. Heavy rare gas scattering
from these surfaces does not yield rainbow scattering.

Structural information can be derived from atom-
solid scattering but further advances appear to depend
upon the development of data which includes energy
analysis of the scattered beam and the use of cryogenic
crystal surfaces.

References

1. For a more general review see: Saltsburg, H. Ann.
    Rev. Phys. Chem., 24, 1973

2. Stickney, R. E. Advan. At. Mol. Phys. 3, 1967

3. Pauly, H., Toennies, J. P., Adv. At. Mol. Phys. 1,
    1965

4. Anderson, J. B., Andres, R. P., Fenn, J. B., Advan.
    Chem. Phys. 10, 275 (1966)

5. Yamamoto, S. and Stickney, R. E., J. Chem. Phys.
    47, 1091 (1967)

6. Smith, J. N., Saltsburg, H., in Fundamentals of Gas
    Surface Interactions, Academic Press (1967)

7. Estermann, I., Stern, O., Z. Phys. 61, 95 (1930)

8. Hoinkes, H., Nahr, H., Wilsch, H., Surface Sci. 30,
    363 (1972)

9. O'Keefe, D. R., Palmer, R. L., Saltsburg, H., Smith,
    J. N., J. Chem. Phys., 49, 5194 (1968)

10. O'Keefe, D. R., Palmer, R. L., Saltsburg, H., Smith,
    J. N., Surf. Sci., 20, 27 (1970)

11. Tendulkar, D. V., Stickney, R. E., ibid., 27, 516
    (1971)

12. Ehrlich, G., Proc. Third Intnl. Congress on Catalysis
    (North Holland, 1965) vol. 1, p. 125

13. Sikafus, E. N., Bonzel, H. P., in Progress in Surface
    and Membrane Science, 4, 115 (1971)

14. Weinberg, W. H. and Merrill, R. P., J. Chem. Phys.,
    56, 2893 (1972)

15. Weinberg, W. H., Merrill, R. P., in Adsorption-Desorp-
    tion Phenomena, (Academic Press, 1972), p. 151

16. Lennard Jones, J. E., Devonshire, A. F., Nature, 137,
    1069 (1936)

17. Hoinkes, H., Nahr, H., Wilsch, H., J. Phys. C5, 1432
    (1972)

18. Daunt, J. G. and Lerner, E. in Ref. 15, p. 127

19. Williams, B. R., J. Chem. Phys., 55, 3220 (1971)

20. Yamamoto, J., Stickney, R. E., ibid., 53, 1594 (1970)

21. Mason, B. F., Williams, B. R., ibid, 56, 1895 (1972)

22. Palmer, R. L., O'Keefe, D. R., Saltsburg, H., Smith,
    J. N., J. Vac. Sci. Tech. 7, 91 (1970)

23. Mason, B. F., Williams, B. R., J. Chem. Phys., 57,
    872 (1972)

24. McClure J. D., ibid, 57, 2810, 2833 (1972)

25. O'Keefe, D. R., Palmer, R. L., Smith, J. N., ibid,
    55, 4572 (1971)

26. McClure, J. D., ibid, <u>52</u>, 2712 (1970)

27. Lorenzen, J., Raff, L., ibid, <u>52</u>, 1133, 6134 (1970)

28. Fisher, B. B., McMillan, W. G., <u>J. Am. Chem. Soc.</u>, <u>79</u>, 2969 (1957)

ELECTRONIC CHARACTERIZATION OF SUBMONOLAYER FILMS

E. W. Plummer*

National Bureau of Standards

Washington, D.C. 20234

The complete characterization of a surface or a submonolayer film on that surface must specify: (i) the chemical identity of the atoms present; (ii) the geometrical or structural arrangement of these atoms; and (iii) the distribution of electrons around these atoms, both in energy and space.[1,2] At present Auger electron spectroscopy is being used very successfully to identify the atoms present on a surface.[1,3] Low-energy electron diffraction now seems to be at the stage of development theoretically where it can be used to determine the position of atoms on a surface when they are in an ordered array.[4] These techniques furnish very little information about the "chemistry" of the surface. Auger electron spectroscopy looks at the characteristic two electron decay to a core level of the atom and the theory of low-energy electron diffraction has been successful because the incoming electron scatters from the ion core, not the valence electrons.[4] If one is going to investigate the interactions of adsorbed atoms or molecules with each other and the substrate, then it would seem to be essential to use experimental techniques which are sensitive to changes in the electronic structure in the outer-most or valence electrons.

By definition for any form of spectroscopy to be useful for this problem, it must be specific to the surface. Various forms of electron emission spectroscopy are quite sensitive to the surface region, yet each technique has a different sensitivity and produces a different perturbation of the surface or adsorbed film. Therefore, it is advantageous to use several methods. Three methods of probing the electronic structure of the surface and submonolayer films will be described below. The advantages and disadvantages of each technique will be discussed briefly. These

*Present address: University of Pennsylvania, Philadelphia.

three techniques are: ion neutralization spectroscopy (INS),
field emission spectroscopy (FES), and uv photoemission spectroscopy (UPS).

Ion neutralization spectroscopy (INS) has been pioneered and
developed almost solely by Hagstrum.[2] The process occurs when a
slowly moving, singly-charged, positive ion which is near a surface is neutralized to the ground state. This neutralization
occurs via an Auger-like process. One electron from the solid
surface tunnels through the barrier between the surface and the
ion and drops into the ground atomic state. A second electron is
excited by the energy lost by the first. If the excitation energy
is large enough, this second electron may escape from the solid.
In an INS experiment these ejected electrons are energy analyzed
and then the distribution is unfolded to obtain the "one electron
density of states at the surface." There are several specific
studies using this technique, which are referenced in the article
by Hagstrum.[2] This technique is surface sensitive because the
neutralizing electron must tunnel from the solid, and only the
states near the surface will contribute to this process. INS has
the disadvantage that it is a two electron process, requiring an
unfold of the raw data.

Field emission energy distribution measurements have proven
to be very successful for a limited set of applications.[5] This
technique has the great advantage of being very surface sensitive.
All observed electrons must tunnel through a barrier created by
the application of a large electrostatic field, $\sim$ 30 million
volts/cm. A paper by Duke and Alferieff[6] created the present
interest in this field. They showed theoretically that the tunneling of electrons through an adsorbed atom underwent a resonance
effect when the energy of the electron was nearly the same as the
binding energy of the adsorbed atom. This tunneling resonance
effect created structure in the energy distribution of the field
emitted electrons which, when properly analyzed,[5] yielded the
local density of states at the adsorbate.[7]

FES has been used to measure the energy level spectrum of a
wide variety of systems. For example, the spectrum of single
atoms of Ba, Sr, and Ca were observed on single crystal planes of
tungsten[8]; H_2 adsorption on (100) W was monitored as a function of
coverage. In this case, the changes in the energy level spectrum
were observed as the submonolayer film of H_2 went through an order-
disorder transition.[9]

A tunneling electron may also excite the adsorbed specie,
losing energy characteristics of the excited state. In the study
of H_2 adsorption,[9] the vibrational levels of H and D with the
substrate were observed as well as the H-H and D-D vibrational

levels when they were adsorbed as molecules.

A recent theoretical development by Penn and Plummer[10] has shown that the field emission energy distribution from a clean surface measures the one-dimensional local density of states at that surface. We are now in a position to determine how much different the surface and bulk density of states are, if we had a bulk measurement.

The disadvantage of field emission is that the equipment is relatively complicated and is not available commercially. In addition, it has a limited energy range below the Fermi energy, approximately 2-3 eV.

Ultraviolet photoemission (UPS) shows the greatest promise of these three techniques, primarily because of the ease of application and the possibility of excitation from states lying in a wide energy range. Electrons are excited by absorption of a photon. The excited electrons which escape are then energy analyzed. UPS is surface sensitive because the escape depth of the excited electron is very short, 4-10 A.[4] If there were not any matrix element effects or final state effects, then the energy distribution would yield the initial density of states. Several problems are undoubtedly going to develop as more experimental data becomes available:

(1) There will be strong matrix elements.

(2) Solids have a very pronounced final state band structure which in general is not free electron like.

(3) UPS is only partially surface sensitive, maybe 30% of the signal is from the surface. Therefore, the observed energy distribution is a mixture of bulk and surface effects, making it difficult to clearly measure either accurately.

The practical solution to these problems, before an adequate theory is developed, is to compare UPS with FES and INS. INS and especially FES are on much better theoretical footing. Hagstrum has compared INS and UPS for adsorption on Ni surfaces,[11] and we have compared FES and UPS for clean single crystal faces of W[10] and adsorption on W.[12]

The power of this technique can be illustrated by a few examples. Eastman and Cashion showed that CO was not dissociated upon adsorption on nickel.[13] We have shown it is dissociated on one plane of tungsten but not on another.[14] CO on tungsten goes through irreversible conversions upon heating or changing the coverage which could not be observed by many other techniques since nothing is desorbed from the surface in the process of conversion.

References:

1. C. B. Duke and R. L. Park, Physics Today 25, 23 (Aug. 1972).
2. H. D. Hagstrum, Science 178, 275 (Oct. 1972).
3. P. W. Palmberg, in Electron Spectroscopy, D. A. Shirley, ed. (North-Holland, Amsterdam).
4. C. B. Duke (paper presented at this meeting).
5. J. W. Gadzuk and E. W. Plummer, Rev. Mod. Phys. (July 1973).
6. C. B. Duke and M. E. Alferieff, J. Chem. Phys. 46, 926 (1967).
7. D. Penn, R. Gomer and M. H. Cohen, Phys. Rev. Letters 27, 26 (1971); Phys. Rev. B5, 768 (1972).
8. E. W. Plummer and R. D. Young, Phys. Rev. B1, 2088 (1970).
9. E. W. Plummer and A. E. Bell, J. Vac. Sci. Tech. 9, 583 (1972).
10. D. R. Penn and E. W. Plummer, submitted to Phys. Rev.
11. H. D. Hagstrum, J. Vac. Sci. Tech. 10, 264 (1973).
12. B. J. Waclawski and E. W. Plummer, Phys. Rev. Letters 29, 783 (1972).
13. D. E. Eastman and J. K. Cashion, Phys. Rev. Letters 27, 1520 (1971).
14. E. W. Plummer and B. J. Waclawski (in preparation).